Intervirology

International Journal of Basic and Medical Virology

Founded 1973 by J.L. Melnick
Continued by F. Rapp (1986–1990); M.J. Buchmeier and C.R. Howard (1991–1993)

KARGER

Printed in Switzerland
on acid-free paper by
Thür AG Offsetdruck, Pratteln

Appears bimonthly:
1 volume per year (6 issues)

Intervirology

Submission: Only original papers written in English are considered and should be sent to:

S. Karger AG
Editorial Dept. 'Intervirology'
P.O. Box
CH–4009 Basel (Switzerland)

Manuscripts should be submitted fourfold, i.e., one original and three copies (with four sets of illustrations of which one is an original), typewritten double-spaced on one side of the paper, with a wide margin.

Conditions: All manuscripts are subject to editorial review. Manuscripts are received with the explicit understanding that they are not under simultaneous consideration by any other publication. Submission of an article for publication implies the transfer of the copyright from the author to the publisher upon acceptance. Accepted papers become the permanent property of 'Intervirology' and may not be reproduced by any means, in whole or in part, without the written consent of the publisher. It is the author's responsibility to obtain permission to reproduce illustrations, tables, etc. from other publications.

Arrangement:
Title page: The first page of each paper should indicate the title (main title underlined), the authors' names, and the institute where the work was conducted. A short title for use as running head as well as the full address of the author to whom correspondence should be sent are also required.
Full address: The exact postal address complete with postal code must be given at the bottom of the title page. Please also supply Phone, Fax, and E-mail numbers.
Key words: For indexing purposes, a list of 3–10 key words in English is essential.
Summary: Each paper needs a summary of up to 10 lines.
Footnotes: Avoid footnotes. When essential, they are numbered consecutively and typed at the foot of the appropriate page.
Body of the paper: Manuscripts should then be divided into the following subdivisions: Introduction, Materials and Methods (as brief as possible), Results, Discussion, Acknowledgments and References.
Tables and illustrations: Tables and illustrations (both numbered in Arabic numerals) should be prepared on separate sheets. Tables require a heading and figures a legend, also prepared on a separate sheet. For the reproduction of illustrations, only good drawings and original photographs can be accepted; negatives or photocopies cannot be used. When possible, group several illustrations on one block for reproduction (max. size 181 × 223 mm) or provide crop marks. On the back of each illustration, indicate its number, the author's name, and 'top'.

Color illustrations: Up to 6 color illustrations per page can now be integrated with the text at the reduced price of SFr. 600.–/US$ 400.00 per page. Color illustrations are reproduced at the author's expense.

References: In the text identify references by Arabic numerals [in square brackets]. Material submitted for publication but not yet accepted should be noted as 'unpublished data' and not be included in the reference list. The list of references should include only those publications which are cited in the text. Do not alphabetize; number references in the order in which they are first mentioned in the text. The surnames of the authors followed by initials should be given. There should be no punctuation other than a comma to separate the authors. Cite all authors, 'et al' is not sufficient. Abbreviate journal names according to the Index Medicus system. (Also see International Committee of Medical Journal Editors: Uniform requirements for manuscripts submitted to biomedical journals. Br Med J 1988; 296:401–405.
Examples
(a) Papers published in periodicals: Kauffman HF, van der Heide S, Beaumont F, Blok H, de Vries K: Class-specific antibody determination against *Aspergillus fumigatus* by means of the enzyme-linked immunosorbent assay. III. Comparative study: IgG, IgA, IgM ELISA titers, precipitating antibodies and IgE binding after fractionation of the anytigen. Int Arch Allergy Appl Immunol 1986;80:300–306.
(b) Monographs: Matthews DE, Farewell VT: Using and Understanding Medical Statistics. Basel, Karger, 1985.
(c) Edited books: Hardy WD Jr, Essex M: FeLV-induced feline acquired immune deficiency syndrome: A model for human AIDS; in Klein E (ed): Acquired Immunodeficiency Syndrome. Prog Allergy. Basel, Karger, 1986, vol 37, pp 353–376.
Further details: Authors may consult our leaflet 'Rules for the Preparation of Manuscripts', available on request from the Publisher, S. Karger AG, P.O. Box, CH–4009 Basel (Switzerland).

Nomenclature of viruses: Names used for viruses should be those approved by the International Committee on Taxonomy of Viruses (ICTV) and published in the fourth report of the ICTV 'Classification and Nomenclature of Viruses', Intervirology *17:* 1–200 (1982). If desired, synonyms may be added in parentheses when the name is first mentioned. Approved generic (or group) and family names should also be used.

Short communications: In addition to regular papers, a section of the journal will publish brief communications of important findings, in which illustrative material should be limited. Such notes should include a brief summary and key words, the body of the report (without any subdivisions), brief acknowledgment, and a few selected references. The general style and submission procedures outlined above apply to the short communication.

Commentaries: Short commentaries of an editorial nature may be submitted to the journal. In these communications, usual manuscript subdivisions do not apply, and a summary statement is not needed; however, a very brief reference list may be included.

Page charges: There is no page charge for papers of 6 or fewer printed pages (including tables, illustrations and references). Each additional complete or partial page is charged to the author at SFr. 225.–/US $ 150.00. The allotted size of a paper is equal to approx. 16 manuscript pages (including tables, illustrations and references).

Galley proofs: Unless indicated otherwise, galley proofs are sent to the first-named author and should be returned with the least possible delay. Alterations made in galley proofs, other than the correction of printer's errors, are charged to the author. No page proofs are supplied.

Reprints: Order forms and a price list are sent to the author by the publisher after the article has been set in type. Orders submitted after the issue is printed are subject to considerably higher prices. Allow five weeks from date of publication for delivery of reprints.

KARGER

Standard forms of abbreviations: The standard abbreviations used for measurements of weight, length, volume, and time (i.e., mg, cm, ml, and min) are acceptable and may be used without definition. The abbreviations listed below also require no preliminary definition. Please note that punctuation is not needed. When using other abbreviations, the name should be spelled out in full at the first mention, and followed with the abbreviation in parentheses.

°	degree centigrade (Fahrenheit will not be used)
A, C, G, T, U	adenine, cytosine, guanine, thymine, uracil
AMP	5'-monophosphates of adenosine
CMP	– cytidine
GMP	– guanosine
TMP	– thymidine
UMP	– uridine
ADP, etc.	adenosine 5'-diphosphate, etc.
ATP, etc.	adenosine 5'-triphosphate, etc.
CPE	cytopathic effect
DEAE-cellulose	diethylaminoethylcellulose
DNA	deoxyribonucleic acid
DNase	deoxyribonuclease
EDTA	ethylenediaminetetra-acetate
EOP	efficiency of plating
g	acceleration of gravity
ID_{50}	median infective dose
LD_{50}	median lethal dose
M	molar
MEM	minimum essential medium
MOI	multiplicity of infection
N	normal
ng	nanogram
nm	nanometer (10^{-9} m)
PAGE	polyacrylamide gel electrophoresis
PFU	plaque forming units
RNA	ribonucleic acid
RNase	ribonuclease
rpm	revolutions per minute
SD	standard deviation
SDS	sodium dodecyl sulfate
sp gr	specific gravity
TCD_{50}	median tissue culture infective dose
Tris	tris (hydroxymethyl) aminomethane (2-amino-2-hydroxymethylpropane-1,3-diol)
UV	ultraviolet
V	volt
W	watt

KARGER

General Information

Intervirology

Publication data: 'Intervirology' is published 6 times annually. Volume 37 with 6 issues appears in 1994.

Subscription rates: Subscriptions run for a full calendar year. Prices are given per volume.

Personal subscription:

Subscription + postage and handling
SFr. 453.60 + SFr. 16.80 Europe
SFr. 453.60 + SFr. 25.20 Overseas
US$ 303.10 + US$ 16.80
(Must be in the name of, billed to, and paid by an individual. Order must be marked 'personal subscription'.)

Institutional subscription:

Subscription + postage and handling
SFr. 648.– + SFr. 24.– Europe
SFr. 648.– + SFr. 36.– Overseas
US$ 433.00 + US$ 24.00

Airmail postage: SFr. 36.–, US$ 30.00. (Extra per volume.)

Discount subscription prices:
International Union of Microbiological Societies, American Association of Immunologists.

Advertising: Correspondence and rate requests should be addressed to the Publisher.

Microforms: Please direct all enquiries and orders to:
University Microfilms Inc.
300 North Zeeb Road, P.O. Box 1346
Ann Arbor, MI 48106-1346 (USA)

Subscription orders: Orders can be placed at agencies, bookstores, directly with the Publisher S. Karger AG, P.O. Box, CH–4009 Basel (Switzerland)
or with any of the following distributors:

Deutschland: S. Karger GmbH, Postfach
D–79095 Freiburg

France: Librairie Luginbühl
36, bd de Latour-Maubourg, F–75007 Paris

Great Britain, Ireland: S. Karger Publishers
11 Ravenscroft Road
Beckenham, Kent BR3 4TP

USA: S. Karger Publishers, Inc.
26 West Avon Road, P.O. Box 529
Farmington, CT 06085

India, Bangladesh, Sri Lanka:
S. Karger India Subscription Agency
B–5/132 Safdarjung Enclave
New Delhi-110029

Japan: Katakura Libri, Inc.
36-9 Hongo 3-chome, Bunkyo-ku, Tokyo 113

Change of address: Both old and new address should be sent to the subscription source.

Single issues and back volumes: Information can be obtained through the Publisher.

Bibliographic indices: This journal is regularly listed in bibliographic services, including *Current Contents®*.

KARGER

Hard cover reprint

Reprint of Intervirology, Vol. 35, No. 1–4 (1993)
Editors: M.J. Buchmeier, La Jolla, Calif.; C.R. Howard, London

Viruses and Virus-Like Agents in Disease

A Karger Symposium
Basel, March 7–9, 1993

Editors:

Rolf M. Zinkernagel, Zürich

Werner Stauffacher, Basel

This publication is dedicated to the question of how viruses or virus-like agents may be involved in the etiology of a number of chronic and degenerative diseases, whose pathogenesis is still not clearly understood. Renowned specialists in basic and clinical virology and immunology discuss topics such as hepatitis, various autoimmune diseases, heart disease, and spongiform encephalopathies.

Contents

Zinkernagel, R.M.: Preface · *Stauffacher, W.:* Welcome Address

Introductory Lecture

Koprowski, H.: Yesterday, Today and Tomorrow in Virology

Virus Detection, Viral Ecology and Disease

Webster, R.G.; Wright, S.M.; Castrucci, M.R. et al.: Influenza - A Model of an Emerging Virus Disease · *Wolf, H.; Bogedain, C.; Schwarzmann, F.:* Epstein-Barr Virus and Its Interaction with the Host · *Blum, H.E.:* Hepatitis B Virus: Significance of Naturally Occurring Mutants · *Brownlie, J.; Clarke, M.C.:* Experimental and Spontaneous Mucosal Disease of Cattle: A Validation of Koch's Postulates in the Definition of Pathogenesis · *Jahn, G.; Plachter, B.:* Diagnostics of Persistent Viruses: Human Cytomegalovirus as an Example

Virus-Induced Autoimmune Diseases

Imai, H.; Kiyosawa, K.; Chan, E.K.L. et al.: Autoantibodies in Viral Hepatocellular Carcinoma · *ter Meulen, V.; Liebert, U.G.:* Measles Virus-Induced Autoimmune Reactions against Brain Antigen · *Wekerle, H.:* T-Cell Autoimmunity in the Central Nervous System · *Wick, G.; Trieb, K.; Aguzzi, A. et al.:* Possible Role of Human Foamy Virus in Graves' Disease · *Manns, M.P.:* Viruses in Autoimmune Liver Disease

Viral Immunopathology

Oldstone, M.B.A.; Rall, G.F.: Mechanisms and Consequence of Viral Persistence in Cells of the Immune System and Neurons · *Eddleston, A.L.W.F.:* Virus- and Immune-Mediated Liver Damage in Hepatitis · *Vallbracht, A.; Fleischer, B.; Busch, F.W.:* Hepatitis A: Hepatotropism and Influence on Myelopoiesis · *Kandolf, R.; Klingel, K.; Zell, R. et al:* Molecular Pathogenesis of Enterovirus-Induced Myocarditis: Virus Persistence and Chronic Inflammation · *Stevens, J.G.:* HSV-1 Neuroinvasiveness

'Viruses Seeking Diseases/ Diseases Seeking Viruses'

Weissmann, C.; Büeler, H.; Fischer, M. et al.: Role of the PrP Gene in Transmissible Spongiform Encephalopathies · *Kalden, J.R.; Herrmann, M.:* Autoimmune Diseases in Humans, e.g. Autoimmune Rheumatic Diseases · *Diener, T.O.; Owens, R.A.; Hammond, R.W. et al.:* Viroids: The Smallest and Simplest Agents of Infectious Diseases. How Do They Make Plants Sick? · *Neumann-Haefelin, D.; Fleps, U.; Renne, R. et al.:* Foamy Viruses · *Kimberlin, R.H.:* Bovine Spongiform Encephalopathy: An Appraisal of the Current Epidemic in the United Kingdom

Fields of Interest

Virology, Infectious Diseases, Immunology, Immunopathology, Internal Medicine, Microbiology

Viruses and Virus-Like Agents in Disease
A Karger Symposium
Basel, March 7–9, 1993
Reprint of 'Intervirology', Vol. 35, No. 1–4 (1993)
Editors: Zinkernagel, R.M. (Zürich); Stauffacher, W. (Basel)
220 p., 71 fig., 39 tab., hard cover, 1993
SFr. 179.– / DM 214.– / US$ 143.25
Prices subject to change
DM price for Germany, US$ price for USA only
ISBN 3-8055-5785-X

KI 93066

KARGER

Papillomaviruses

Guest Editor
L. Gissmann, Maywood, Ill.

14 figures and 9 tables, 1994

KARGER

S. Karger
Medical and Scientific Publishers
Basel · Freiburg · Paris · London
New York · New Delhi · Bangkok
Singapore · Tokyo · Sydney

Drug Dosage
The authors and the publisher have exerted every effort to ensure that drug selection and dosage set forth in this text are in accord with current recommendations and practice at the time of publication. However, in view of ongoing research, changes in government regulations, and the constant flow of information relating to drug therapy and drug reactions, the reader is urged to check the package insert for each drug for any change in indications and dosage and for added warnings and precautions. This is particularly important when the recommended agent is a new and/or infrequently employed drug.

Printed in Switzerland on acid-free paper by
Thür AG Offsetdruck, Pratteln
ISBN 3–8055–6096–6

Contents

Gissmann, L.	Editorial	141
Pfister, H.; Fuchs, P.G.	Anatomy, Taxonomy and Evolution of Papillomaviruses	143
Chow, L.T.; Broker, T.R.	Papillomavirus DNA Replication	150
Fuchs, P.G.; Pfister, H.	Transcription of Papillomavirus Genomes	159
Stöppler, H.; Conrad Stöppler, M.; Schlegel, R.	Transforming Proteins of the Papillomaviruses	168
Altmann, A.; Jochmus, I.; Rösl, F.	Intra- and Extracellular Control Mechanisms of Human Papillomavirus Infection	180
Brandsma, J.L.	Animal Models of Human-Papillomavirus-Associated Oncogenesis	189
Schneider, A.	Natural History of Genital Papillomavirus Infections	201
Gross Fisher, S.	Epidemiology: A Tool for the Study of Human Papillomavirus-Related Carcinogenesis	215
Cowsert, L.M.	Treatment of Papillomavirus Infections: Recent Practice and Future Approaches	226
	Author Index	231
	Subject Index	232

Intervirology 1994;37:141–142

L. Gissmann

Department of Obstetrics and Gynecology,
Loyola University Medical Center,
Maywood, Ill., USA

Editorial

Research on papillomaviruses has travelled a long and winding road. To date, our growing awareness of the biology of these viruses has so far culminated in the definition of their critical role in the evolution of anogenital cancer in humans. Investigators are now considering the development of antiviral vaccines or other virus-specific strategies in order to interfere with this type of malignancy [for a review see ref. 1; Cowsert, this volume].

Several discoveries during the past 90 years form the basis of our current knowledge of papillomaviruses. The infectious nature of warts was established at the turn of this century and important observations such as the host specificity of human and animal papillomaviruses was discovered in the early days [for a review see ref. 2]. In the twenties, the principles of progression of benign papillomas into malignant cancer expedited by the synergism of chemical carcinogens were first described in the cottontail rabbit papillomavirus system. With the advent of modern technologies this and other animal models are still of great value for the study of virus-host interactions [see Brandsma, this volume]. The physicochemical characterization of virus particles and their DNA was initiated in the 1960s when virions were purified from skin warts in which efficient replication takes place [see Pfister and Fuchs, this volume]. It became apparent during this period that papillomaviruses stubbornly evade the investigators' curiosity by not replicating in cell culture. Since negative results often remain unpublished, it is unclear how many years of frustrating work have actually been spent at laminar flow benches before researchers reached the disappointing conclusion that the differentiation of epithelium, apparently critical for papillomavirus replication, cannot be fully imitated in culture. It was a milestone in papillomavirus research when Kreider and colleagues succeeded in propagating HPV type 11 particles after grafting infected human tissue into nude mice [review by Brandsma, this volume]. As elegant and informative as this system clearly is, it comprises labor-intensive and sophisticated experiments and obviously does not permit the production of infectious particles at a preparative scale. The more recently introduced epithelial raft system may finally help to overcome what is still a major deficit in papillomavirus research [for a review see ref. 3; Chow and Broker, this volume]. Another promising avenue has been opened by the production of papillomavirus-like particles after expression of the structural proteins in recombinant vectors such as vaccinia [for a review see ref. 4; Pfister and Fuchs, this volume]. It can be expected that complete (i.e. DNA-containing and hopefully infectious) virions will soon become available by the aid of such systems.

Probably more than with any other virus systems, the advance of recombinant DNA technology in the 1970s boosted papillomavirus research. Investigation of the genome organization, the regulation of gene expression and the characterization of their products has resulted in the accumulation of a bulk of interesting information fostering our understanding of mechanisms of HPV-related carcinogenesis [see Fuchs and Pfister; Stöppler et al., this volume]. It turned out that the genomes of different papillomaviruses are similar in their organization, thus permitting cross-hybridization at nonstringent conditions [5; Pfister and Fuchs, this volume]. In fact, many of the more than 70 HPV types have been identified after direct cloning of the DNA from infected tissues by using well-defined genomes or, more recently, general PCR primers

L. Gissmann
Department of Obstetrics and Gynecology
Division of Research, Loyola University Medical Center
2160 South First Avenue
Maywood, IL 60153 (USA)

0300–5526/94/0374–0141
$8.00/0

as probes for nonstringent hybridization. This strategy permitted the detection of new viruses without prior purification of virus particles, which in most cases replicate at an extremely low level in vivo.

What have we learned by identifying so many HPV types? I remember that some time ago it was suggested that we should restrict ourselves to the characterization of one or a few papillomaviruses as was previously done when investigating the genetics of *Escherichia coli* bacteriophages. The argument raised against this position was that papillomaviruses are human pathogens and that we must first identify the most relevant agents before deciding which one(s) need(s) to be studied in detail. Following this advice led to the discovery of HPV 16 and other viruses related to the development of cancer of the anogenital and (as more recently found) respiratory tract. In fact, these viruses are now being studied by many laboratories [for reviews see ref. 6, 7]. In addition, DNA sequence analysis of a number of individual human and animal papillomaviruses and of intratype HPV variants isolated from different geographic regions is a powerful tool for studying the evolution of these viruses and their hosts [see Pfister and Fuchs, this volume].

Despite the firmly established role of HPV in human cancer and the growing understanding of mechanisms of cell transformation, many aspects of the biology of papillomaviruses remain enigmatic. Some of these aspects have been addressed in this volume: what is the natural history of papillomavirus infections; what can we expect from future epidemiologic studies; how do papillomaviruses manage to replicate in terminally differentiated cells; what are the prospects for anti-viral therapies [see contributions by Schneider, Cowsert, Fisher, Chow and Broker]? The growing area of research on the immune biology of HPV infections as well as the anticipation of vaccine developments have recently been reviewed [1, 8–10]. The article by Altmann et al. discusses immunological aspects specific to the control of HPV gene expression via cell-cell interactions.

I am grateful to my colleagues who contributed to this volume, which is intended to provide a snapshot of a rapidly moving field.

References

1 Campo MS: Vaccination against papillomavirus in cattle. Curr Top Microbiol Immunol 1994;186:255–266.

2 Rowson KE, Mahy BW: Human papova (wart) virus. Bact Rev 1967;31:110–131.

3 Meyers C, Laimins L: In vitro systems for the study and propagation of human papillomaviruses. Curr Top Microbiol Immunol 1994;186: 199–215.

4 Hagensee ME, Galloway DA: Growing human papillomaviruses and virus-like particles in the laboratory. Papillomavirus Rep 1993;4:121–124.

5 Law MF, Lancaster WD, Howley PM: Conserved polynucleotide sequences among the genomes of papillomaviruses. J Virol 1979;32: 199–207.

6 zur Hausen H: Viruses in human cancers. Science 1991;254:1167–1173.

7 Snijders PJF, van den Brule AJC, Meijer CJLM, Walboomers JMM: Papillomaviruses and cancer of the upper digestive and respiratory tracts. Curr Top Microbiol Immunol 1994; 186:177–198.

8 Frazer I, Tindle B: Immune response to human papillomaviruses and the prospects for human papillomavirus-specific immunization. Curr Top Microbiol Immunol 1994;186:217–254.

9 Altmann A, Gissmann L, Jochmus I: Towards HPV vaccination; in Minson A, Neil J, McCrae M (eds): Viruses and Cancer. Cambridge, Cambridge University Press, 1994, pp 71–80.

10 Stern P, Stanley M (eds): Human Papillomaviruses and Cervical Cancer. Oxford, Oxford University Press, 1994.

Intervirology 1994;37:143–149

H. Pfister
P.G. Fuchs

Institut für Klinische und Molekulare Virologie der Friedrich-Alexander-Universität Erlangen-Nürnberg, Deutschland

Anatomy, Taxonomy and Evolution of Papillomaviruses

Key Words
Evolution
Gene
Papillomavirinae
Phylogeny
Structure
Taxonomy
Virion
Virus-like particles

Papillomavirinae is a subfamily of the Papovaviridae family [1]. This taxonomic placement is based on the common capsid structure and similar biochemical composition of papillomaviruses (PVs) and polyomaviruses, which form the second subfamily of Papovaviridae. Members of the two subfamilies have different genome organizations, in that there is only one coding DNA strand in PVs whereas separate strands code for vegetative functions and structural proteins of polyomaviruses [2]. The DNAs of viruses from different subfamilies do not hybridize even at low stringency and there is no serological cross-reactivity. Limited homologies in nucleotide and protein-binding domains of viral polypeptides may reflect a distant relationship or functional convergence. From these data, PVs and polyomaviruses split very early in evolution if there was a common ancestor at all.

Virion Structure

The PV virions are nonenveloped particles of 55–60 nm which encapsulate a single copy of the double-stranded, circular DNA genome in the form of a chromatin-like complex with cellular histones [2]. They have a density in cesium chloride of 1.34–1.36 g/l and a sedimentation coefficient ($S_{20,w}$) of 300 [3, 4]. Due to the lack of lipids, the PV particles are resistant to ether and other lipid solvents.

The viral capsid consists of 72 capsomeres, 12 of which contact five neighbors whereas the others are six-coordinated. They are arranged on a skew T = 7d icosahedral surface lattice [5–7]. As shown by cryoelectron microscopy, a capsomere consists of a trunk that broadens both distally and proximally [7]. The capsomeres associate at the base to form an approximately 2-nm thick shell layer and project radially to a height of about 5.8 nm, ending in a five-pointed, star-shaped head. A cylindrical channel of approximately 2.8 nm can be recognized centrally within each capsomere, extending along its axis from the interior of the particle to a point half way to the capsomere surface. As demonstrated for bovine (B) PV 1 and human (H) PV 1, all capsomeres display a 5-fold symmetry [7], and are therefore most probably pentamers composed of five molecules of the major capsid protein L1, which contributes 80–90% of the protein content of the viral shell [8, 9].

The L1 proteins have an average molecular mass of 55,000 and their amino acid sequences are highly con-

H. Pfister
Institut für Klinische und Molekulare Virologie
Friedrich-Alexander-Universität Erlangen-Nürnberg
Schlossgarten 4
D-91054 Erlangen (Germany)

0300–5526/94/0374–0143
$8.00/0

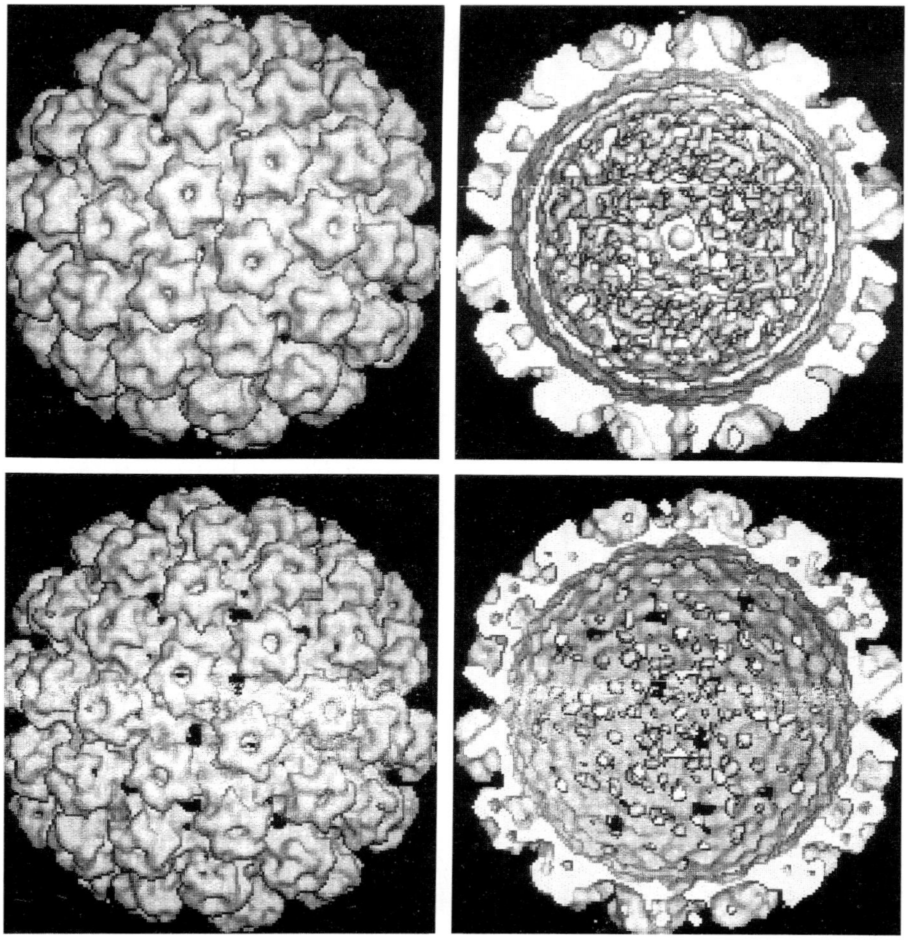

Fig. 1. Cryoelectron microscopy modelling of HPV 1 capsids from warts (top) or VLPs (bottom). Outside view to the left and inside view to the right. (Photos kindly provided by M.E. Hagensee, N.H. Olson, T.S. Baker and D.A. Gallaway.)

served among all PVs [10]. The analysis of BPV 1 particles by two-dimensional gel electrophoresis and isoelectric focusing revealed an amazing variety among the L1 molecules with the same molecular weight [11].

The L1 protein from BPV 1 particles and the HPV 16 L1 expressed by a recombinant vaccinia virus were shown to be partially glycosylated [11–13]. L1 glycosylation is somewhat puzzling since the process is generally believed to occur via the secretive protein pathway targeting the modified proteins to the cell surface and not to the nucleus where the assembly of PV capsids takes place. Recent data of Zhou et al. [14] showed the glycosylated HPV 16 L1 molecules as being 'trapped' within the endoplasmic reticulum and exported neither to the surface membrane nor to the nucleus. Further studies are necessary to elucidate the role of the L1 modifications in the viral life cycle and pathogenicity.

A second, minor component of the PV particles is the viral L2 protein. The L2 polypeptides with an apparent M_r of about 76 kD [9, 15] are less well conserved than the L1 proteins and their function(s) remains so far rather obscure. Capsids containing only L1 (expressed from a recombinant vaccinia virus) appear identical to capsids containing both L1 and L2 proteins [16]. The amino-terminus of L2 is interestingly able to bind DNA in a sequence-unspecific manner [17]. This observation may suggest a role for L2 either in attracting viral DNA during ssembly of viral particles or in proper positioning of the minichromosome within the capsid.

The capsid proteins carry type-specific and genus-specific antigenic determinants. The latter appear to be largely masked in intact particles, but can be displayed with unassembled polypeptides and cross-reacting antiserum raised against detergent-disrupted virions [18].

Studies involving the PV particles were generally hampered by difficult access to sufficient amounts of native virions. PV capsids are formed in infected keratinocytes only during the very last stages of their differentiation and

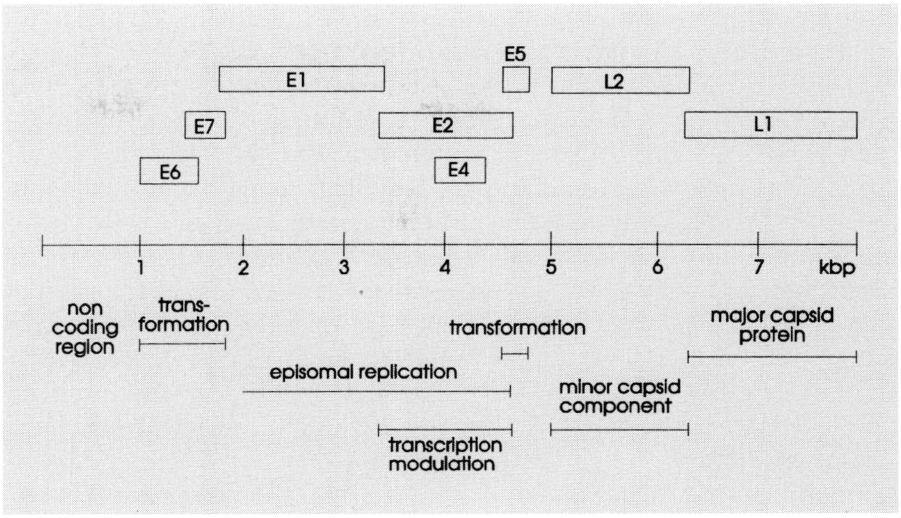

Fig. 2. Organization of a PV consensus genome. The genome is marked off in kilobase pairs (kbp). The open bars represent open reading frames labelled 'E' or 'L' depending on their position in the 'early' or 'late' region of the genome. Gene functions are listed below the genome.

the amount of viral particles in different PV-induced tumors may differ substantially [19]. Moreover, there is still no generally applicable productive in vitro system for PVs.

A new technical approach to studies of PV particles has been introduced by Zhou et al. [20]. They demonstrated the ability of the HPV 16 L1 and L2 proteins expressed simultaneously by means of a recombinant vaccinia virus vector to assemble into empty virus-like particles (VLPs) in the nuclei of the infected cells. As shown by subsequent studies with HPV 1, 6, 11, 16, BPV 1 and cottontail rabbit (CR) PV, the in vivo or in vitro potential to form VLPs was in fact an intrinsic property of the L1 proteins whereas L2 on its own failed to self-assemble [21–25]. However, the presence of L2 significantly increased the efficiency of the L1 assembly process [22, 23, 25]. Interestingly, low yields of VLPs observed for L1 derived from the HPV 16 prototype DNA could be traced back to a single nonconserved amino acid change in its sequence [22]. Most VLPs display almost authentic particle morphology. HPV 1 VLPs were indistinguishable by conventional and cryo-electron microscopy from original empty particles isolated from plantar warts [16, 23] (fig. 1).

There are two properties of the VLPs which may be of practical use: (1) It could be demonstrated that VLPs are highly immunogenic and most probably carry the same conformational epitopes as native particles [22, 24]. Taking into account that VLPs of any PV can be produced at a preparative scale, they offer a unique source of authentic viral antigens. This opens new perspectives in studies of the host immune response to an HPV infection and in designing PV vaccines. (2) As shown so far for BPV1, the VLPs assembled in vivo may package viral DNA and are infectious for susceptible cells [25]. The experiences with other viruses, e.g. polyoma, suggest that packaging of any DNA of proper size into VLPs may also be accomplished in vitro [26, 27]. These properties allow us to envisage PV VLPs as vectors for the delivery of nucleic acids (and probably also other molecules) to their target tissues. More immediately, the VLPs should certainly facilitate the identification of the cellular receptor(s) for PVs.

Genome Organization

The double-stranded circular PV DNAs are composed of 7,500–8,000 base pairs. The organization of a consensus genome is shown in figure 2. All protein-encoding sequences are located on one DNA strand. The individual reading frames are designated 'early' (E) or 'late' (L) to indicate their expression in the viral life cycle. Comprehensive studies of several animal and human PVs have identified the genes whose products are involved in viral DNA replication, transcription control, cell transformation, and capsid assembly (see other articles in this issue). An early function of E4 is not yet clear. Late in the viral life cycle, E4 proteins seem to be involved in the collapse of the cytokeratin network and in virus release [28]. Reading frames L1 and E6 are separated by 400–1,000 base pairs that do not encode proteins ['noncoding region' (NCR)] but contain the origin of viral DNA replication and transcription control signals.

In animal PVs inducing fibropapillomas (BPV1, deer PV, European elk PV) and in a subgroup of HPV (6, 16), reading frame E5 encodes a hydrophobic protein that plays a role in cell transformation [29, 30]. In other PV (BPV 4, HPV 1, 5, and 8) there is no comparable E5 [31–34]. A subgroup of BPVs (types 3, 4, and 6) contain no E6 gene [35]. There is a reading frame E8 at the corresponding position of the genome, which could encode a hydrophobic protein similar to E5. Both genes might be analogous or reflect a genomic rearrangement early in evolution.

Classification

The classification of PVs is based on the host range and the relatedness of the DNAs [36]. Each virus is named first after its natural host (e.g. human, bovine, rhesus monkey), which is usually unequivocal because PVs rarely or never cross the host species barriers [37]. Independent isolates from one species were initially defined as different types on the basis of less than 50% cross-hybridization under stringent conditions and were regarded as subtypes if cross-hybridization exceeded the 50% value. Presently, more than 70 HPV types can be differentiated according to this agreement. As DNA sequences can now be easily determined and are more precise than the results of cross-hybridization experiments, a new definition of genotypes has been agreed upon according to which two types share less than 90% nucleotides in their L1, E6, and E7 sequences [38].

A large number of complete or partial HPV DNA sequences have become available within recent years and have allowed a more refined evaluation of relatedness. Viruses that revealed no cross-hybridization at all under stringent conditions still show more than 50% nucleotide homology within conserved regions of the genome such as reading frames E1, E2, and L1. The noncoding control sequences and other reading frames are more divergent [39]. The closest intertype differences in E1 and L1 segments range from 10 to 17% [40]. In contrast, intratype diversity showed a maximum of 2% in E1/L1 and only 5% in the NCR of more than 200 HPV 16 and 18 isolates [41–43]. Variants of HPV 5 differed by 3.6% in the E7 gene and by a remarkable 11% in the E6 gene, which led to the tentative definition of three subtypes [44, 45]. The gap between inter- and intratype diversity suggests that a PV type although originally arbitrarily defined represents a true biological entity like a species.

Phylogeny

Sequence variants of individual HPV types appear clustered in different geographic regions of the world [41–43, 46]. The analysis of phylogenetic trees of HPV 16, 18, and 45 variants sampled throughout the world suggest that these virus types originated in Africa, where the largest number of different variants and the deepest branches are found as well as the closest relatives between HPV 18 and 45. The existence of European, Asian, and American Indian branches indicates that the viruses spread and coevolved with humankind over the globe [42, 43]. In view of the low divergence in specific areas of the world, the rate of viral evolution seems to be very slow. According to rough estimates, the present intratype diversity may have evolved over a period of more than 200,000 years, whereas closely related types separated several million years ago, i.e. before the evolution of humans [43].

It is interesting to note that three nonhuman primate PVs are close relatives of HPVs, namely rhesus monkey PV (RhPV 1) of HPV 11 and 16 [47], *Colobus guereza* monkey PV 1 (CgPV 1) of HPV 3 and 10 [40], and pygmy chimpanzee PV (PCPV 1) of HPV 13 [48]. Some of these animal and human viruses are genetically much more closely related than most HPV types are among each other and share pathogenic properties. The most likely explanation is that a common ancestor of e.g. HPV 13 and PCPV 1 infected an ancestor of humans and chimpanzees and started some diversification after separation of the host species. This would imply that PV types with a distinct pathogenesis already existed in ancient times and evolved separately with their host species.

To group the large number of HPV types, phylogenetic trees were constructed on the basis of nucleotide sequences in the E6, E1, or L1 genes [40, 49–52]. All minor branches of the trees essentially confirmed the grouping based on total genomic hybridization [53] (table 1). This suggests that homologies in subgenomic DNA segments are representative of the relationship between whole genomes and is in line with finding individual HPV types in the same relative position on E6, E1, or L1 trees. These data argue against intertype recombination during the evolution of the presently analyzed HPV types.

Several major branches of the phylogenetic trees bear 'cutaneous types', inducing morphologically and histologically different warts. One highly ramified branch contains all HPV types that are characteristic for epidermodysplasia verruciformis (table 1). They are closely related and share a common organization of their regulatory regions [54]. HPV 37 and 49 have so far been found only

Table 1. Grouping of HPV types according to DNA sequence homology and to association with epidermodysplasia verruciformis, cutaneous and mucosal lesions [51–53, 56, 57, 59, 66]

Cutaneous								Mucocutaneous						Mucosal			
epidermodysplasia verruciformis																	
5	9	24	50	4	1	*41*	*48*	2	3	26	54	7	34	*6*	32	**16**	**18**
8	15			60	63			27	10	*51*		40	64	*11*	42	*31*	*39*
12	*17*			65				*57*	28			43		13		*33*	**45**
36	37							61	29	*30*				44		*35*	59
47	38							62		53				55		*52*	
	49									**56**						*58*	
14										*66*							
19																	
20																	
21																	
22																	
23																	
25																	

Virus types frequently detected in cancers are highlighted in bold, those found less frequently are italicized [67, 68].

in a keratoacanthoma and in flat warts of immunosuppressed patients, respectively [55, 56]. Other deep branches contain only a few types each such as HPV 1/63, HPV 4, 60, 65 [57], HPV 41 [58] or HPV 48 [59]. Although the different subgroups appear rather unrelated to one another, they are assumed to form the heterogenous 'cutaneous lineage' of HPVs including animal skin wart viruses like the CRPV [50, 51].

A highly complex branch combines all HPV types infecting the mucosa and a number of predominantly or exclusively cutaneous types. The members of this 'super group' are closely related in terms of overall DNA sequence homology [39], organization of the early-late region transition, i.e. reading frames E5 and L2 [60], and organization of the control region [61, 62]. This branch was differentiated into subgroups comprising 'mucocutaneous' types and representatives implying a low or a high risk of malignant conversion [49], but this seems to be an oversimplification [52]. Minor branches of the mucocutaneous part combine viruses infecting the skin and mucosa, whereas individual types may show a clear preference for the skin (HPV 3, 10) or the mucosa (HPV 30, 40, 56) [38]. High-risk genital HPVs 16 and 18 appear on different branches, whereas closer relatives of HPV 16 (HPV 31, 35) belong to intermediate-risk types [63]. HPV 13 [64] and HPV 32 [65], which induce focal epithelial hyperplasia of the oral cavity, also represent different branches, and are most closely related to viruses found in genital lesions (table 1).

In summary, there is on the whole quite satisfying correlation between the phylogenetic relationship of HPV types, tissue tropism, and pathogenic properties, which is not too surprising in view of no evidence for intertype recombination and the dogma that related genes should encode similar biological activities. However, numerous exceptions indicate that minor genetic differences, whose functional consequences we cannot yet fully appreciate, may account for significant changes in tropism and pathogenicity. The oncogenic potential is particularly hard to predict from the position on phylogenetic tress because types, which are detected in cancers, arise on different major and minor branches (table 1).

References

1 Murphy FA, Kingsbury DW: Virus taxonomy; in Fields BN, Knipe DM (eds): Virology, New York, Raven, 1990, pp 9–35.

2 Pfister H, Fuchs PG: Papillomaviruses: Particles, genome organization and proteins; in Syrjänen K, Gissmann L, Koss LG (eds): Papillomaviruses and Human Disease. Heidelberg, Springer, 1987, pp 1–18.

3 Pfister H, Gissmann L, zur Hausen H: Partial characterization of the proteins of human papilloma viruses (HPV) 1–3. Virology 1977;83: 131–137.

4 Crawford LV, Crawford EM: A comparative study of polyoma and papilloma viruses. Virology 1963;21:258–263.

5 Finch JI, Klug A: The structure of viruses of the papillomapolyoma type. III: Structure of rabbit papilloma virus. J Mol Biol 1965;13:1–12.

6 Klug A, Finch JI: Structure of the papillomapolyoma type. I. Human wart virus. J Mol Biol 1965;11:403–423.

7 Baker TS, Newcomb WW, Olson NH, Cowsert LM, Olson C, Brown JC: Structures of bovine and human papillomaviruses: Analysis by cryoelectron microscopy and three-dimensional image reconstruction. Biophys J 1991;60: 1445–1456.

8 Favre M, Breitburd F, Croissant O, Orth G: Structural polypeptides of rabbit, bovine, and human papilloma viruses. J Virol 1975;15: 1239–1247.

9 Doorbar J, Gallimore PH: Identification of proteins encoded by the L1 and L2 open reading frames of human papillomaviruses 1a. J Virol 1987;61:2793–2799.

10 Iftner T: Papillomavirus genomes: Sequence analysis related to functional aspects; in Pfister H (ed): Papillomaviruses and Human Cancer, Boca Raton, CRC, 1990, pp 181–202.

11 Larsen PM, Storgaard L, Fey SJ: Proteins present in bovine papillomavirus particles. J Virol 1987;61:3596–3601.

12 Browne HM, Churcher MJ, Stanley MA, Smith GL, Minson AC: Analysis of the L1 gene product of human papillomavirus type 16 by expression in a vaccinia virus recombinant. J Gen Virol 1988;69:1263–1273.

13 Zhou J, Crawford L, McLean L, Sun XY, Stanley M, Almond N, Smith GL: Increased antibody responses to human papillomavirus type 16 L1 protein expressed by recombinant vaccinia virus lacking serine protease inhibitor genes. J Gen Virol 1990;71:2185–2190.

14 Zhou J, Sun XY, Frazer IH: Glycosylation of human papillomavirus type 16 L1 protein. Virology 1993;194:210–218.

15 Komly CA, Breitburd F, Croissant O, Streeck RD: The L2 open reading frame of human papillomavirus type 1a encodes a minor structural protein carrying type specific antigens. J Virol 1986;60:813–816.

16 Hagensee ME, Olson NH, Baker TS, Galloway DA: Three-dimensional structure of vaccinia virus-produced HPV-1 capsids. J Virol, in press.

17 Zhou J, Sun XY, Louis K, Frazer IH: Interaction of human papillomavirus (HPV) type 16 capsid proteins with HPV DNA requires an intact L2 N-terminal sequence. J Virol 1994; 68:619–625.

18 Jenson AB, Rosenthal LD, Olson C, Pass F, Lancaster WD, Shah K: Immunologic relatedness of papillomaviruses from different species. J Natl Cancer Inst 1980;64:495–500.

19 Pfister H: Biology and biochemistry of papillomaviruses. Rev Physiol Biochem Pharmacol 1984;99:111–181.

20 Zhou J, Sun XY, Stenzel DJ, Frazer IH: Expression of vaccinia recombinant HPV16 L1 and L2 ORF proteins in epithelial cells is sufficient for assembly of HPV virion-like particles. Virology 1991;185:251–257.

21 Kirnbauer R, Booy F, Cheng N, Lowy DR, Schiller JT: Papillomavirus L1 major capsid protein self-assembles into virus-like particles that are highly immunogenic. Proc Natl Acad Sci USA 1992;89:12180–12184.

22 Kirnbauer R, Taub J, Greenstone H, Roden R, Dürst M, Gissmann L, Lowy DR, Schiller JT: Efficient self-assembly of human papillomavirus type 16 L1 and L1-L2 into virus-like particles. J Virol 1993;67:6929–6936.

23 Hagensee ME, Yaegashi N, Galloway DA: Self-assembly of human papillomavirus type 1 capsids by expression of the L1 protein alone or by coexpression of the L1 and L2 capsid proteins. J Virol 1993;67:315–322.

24 Rose RC, Bonnez W, Reichman RC, Garcea RL: Expression of human papillomavirus type 11 L1 in insect cells: In vivo and in vitro assembly of viruslike particles. J Virol 1993;67: 1936–1944.

25 Zhou J, Stenzel DJ, Dun XY, Frazer IH: Synthesis and assembly of infectious bovine papillomavirus particles in vitro. J Gen Virol 1993; 74:763–768.

26 Slilaty SN, Berns KI, Aposkian HV: Polyoma-like particle: Characterization of the DNA encapsidated in vitro by polyoma empty capsids. J Biol Chem 1982;257:6571–6575.

27 Silaty SN, Sposkian HV: Gene transfer by polyoma-like particles. Science 1983;220:725–727.

28 Doorbar J, Ely S, Sterling J, McLean C, Crawford L: Specific interaction between HPV-16 E1-E4 and cytokeratins results in collapse of the epithelial cell intermediate filament network. Nature 1991;352:824–827.

29 DiMaio D, Neary K: The genetics of bovine papillomavirus type 1; in Pfister H (ed): Papillomaviruses and Human Cancer. Boca Raton, CRC, 1990, pp 113–144.

30 Banks L, Matlashewski G: Cell transformation and the HPV E5 gene. Papillomavirus Report 1993;4:1–4.

31 Patel KR, Smith KT, Campo MS: The nucleotide sequence and genome organization of bovine papillomavirus type 4. J Gen Virol 1987; 68:2117–2128.

32 Fuchs PG, Iftner T, Weninger J, Pfister H: Epidermodysplasia verruciformis-associated human papillomavirus 8: Genomic sequence and comparative analysis. J Virol 1986;58:626–634.

33 Zachow KR, Ostrow RS, Faras AJ: Nucleotide sequence and genome organization of human papillomavirus type 5. Virology 1987;158: 251–254.

34 Danos O, Katinka M, Yaniv M: Human papillomavirus 1a complete DNA sequence: A novel type of genome organization among papovaviridae. EMBO J 1982;1:231–236.

35 Jackson ME, Pennie WD, McCaffery RE, Smith KT, Grindlay GJ, Campo MS: The B subgroup bovine papillomaviruses lack identifiable E6 open reading frame. Mol Carcinog 1991;4:382–387.

36 Coggin JR, zur Hausen H: Workshop on papillomaviruses and cancer. Cancer Res 1979;39: 545–546.

37 Sundberg JP: Papillomavirus infections in animals; in Syrjänen K, Gissmann L, Koss L (eds): Papillomaviruses and Human Disease. Heidelberg, Springer, 1987, pp 40–103.

38 de Villiers EM: Human pathogenic papillomavirus types: An update; in zur Hausen H (ed): Human Pathogenic Papillomaviruses. Berlin, Springer, 1994, pp 1–12.

39 Pfister H, Krubke J, Dietrich W, Iftner T, Fuchs PG: Classification of the papillomaviruses: Mapping the genome; in Ciba Found Symp 1986;120:3–22.

40 Chan SY, Bernard HU, Ong CK, Chan SP, Hofmann B, Delius H: Phylogenetic analysis of 48 papillomavirus types and 28 subtypes and variants: A showcase for the molecular evolution of DNA viruses. J Virol 1992;66:5714–5725.

41 Chan SY, Ho L, Ong CK, Chow V, Drescher B, Dürst M, ter Meulen J, Villa L, Lunade J, Mgaya HN, Bernard HU: Molecular variants of human papillomavirus type 16 from four continents suggest ancient pandemic spread of the virus and its coevolution with humankind. J Virol 1992;66:2057–2066.

42 Ho L, Chan SY, Burk RD, Das BC, Fujinaga K, Icenogle JP, Kahn T, Kiviat N, Lancaster W, Mavromara-Nazos P, Labropoulou V, Mitrani-Rosenbaum S, Norrild B, Pillai MR, Stoerker J, Syrjaenen K, Syrjaenen S, Tay SK, Villa LL, Wheeler CE, Williamson AL, Bernard HU: The genetic drift of human papillomavirus type 16 is a means of reconstructing pehistoric viral spread and the movement of ancient human populations. J Virol 1993;67:6413–6423.

43 Ong CK, Chan SY, Campo MS, Fujinaga K, Mavromara-Nazos P, Labropoulou V, Pfister H, Tay SK, ter Meulen J, Villa LL, Bernard HU: Evolution of human papillomavirus type 18: An ancient phylogenetic root in Africa and intratype diversity reflect coevolution with human ethnic groups. J Virol 1993;67:6424–6431.

44 Deau MC, Favre M, Orth G: Genetic heterogeneity among human papillomaviruses (HPV) associated with epidermodysplasia verruciformis: Evidence for multiple allelic forms of HPV5 and HPV8 E6 genes. Virology 1991;184: 492–503.
45 Deau MC, Favre M, Jablonska S, Rueda LA, Orth G: Genetic heterogeneity of oncogenic human papillomavirus type 5 (HPV5) and phylogeny of HPV5 variants associated with epidermodysplasia verruciformis. J Clin Microbiol 1993;31:2918–2926.
46 Eschle D, Dürst M, ter Meulen J, Luande J, Eberhardt HC, Pawlita M, Gissmann L: Geographic dependence of sequence variation in the E7 gene of human papillomavirus type 16. J Gen Virol 1992;73:1829–1832.
47 Ostrow RS, LaBresh KV, Faras AJ: Characterization of the complete RhPV 1 genomic sequence and an integration locus from a metastatic tumor. Virology 1991;181:424–429.
48 van Ranst M, Fuse A, Fiten P, Beuken E, Pfister H, Burk RD, Opdenakker G: Human papillomavirus type 13 and pygmy chimpanzee papillomavirus type 1: Comparison of the genome organizations. Virology 1992;190:587–596.
49 van Ranst M, Kaplan JB, Burk RD: Phylogenetic classification of human papillomaviruses: Correlation with clinical manifestations. J Gen Virol 1992;73:2653–2660.
50 van Ranst MA, Tachezy R, Delius H, Burk RD: Taxonomy of the human papillomaviruses. Papillomavirus Rep 1993;4:61–65.
51 Bernard HU, Chan SY, Delius H: Evolution of papillomaviruses; in zur Hausen H (ed): Human Pathogenic Papillomaviruses. Berlin, Springer, 1994, pp 33–54.
52 Bernard HU, Chan SY, Ong CK, Villa LL, Delius H, Peyton CL, Nauer HM, Manos MM, Wheeler CM: Identification and assessment of known and novel human papillomaviruses by polymerase chain reaction, restriction digestion fingerprinting, nucleotide sequence, and phylogenetic algorithms. J Infect Dis, in press.
53 Pfister H: General introduction to papillomaviruses; in Pfister H (ed): Papillomaviruses and Human Cancer. Boca Raton, 1990, pp 1–9.
54 Ensser A, Pfister H: Epidermodysplasia verruciformis associated human papillomaviruses present a subgenus-specific organization of the regulatory genome region. Nucleic Acids Res 1990;18:3919–3922.
55 Scheurlen W, Gissmann L, Gross G, zur Hausen H: Molecular cloning of two new types (HPV37 and HPV38) from keratoacanthoma and malignant melanoma. Int J Cancer 1986; 37:505–510.
56 Favre M, Obalek S, Jablonska S, Orth G: Human papillomavirus type 49, a type isolated from flat warts of renal transplant patient. J Virol 1989;63:4909.
57 Egawa K, Delius H, Matsukura T, Kawashima M, deVilliers EM: Two novel types of human papillomavirus, HPV63 and HPV65: Comparison of their clinical and histological features and DNA sequences to other HPV types. Virology 1993;194:789–799.
58 Hirt L, Hirsch-Behnam A, de Villiers EM: Nucleotide sequence of human papillomavirus (HPV) type 41: An unusual HPV type without a typical E2 binding site consensus sequence. Virus Res 1990;18:179–190.
59 Müller M, Kelly G, Fiedler M, Gissmann L: Human papillomavirus type 48. J Virol 1989; 63:4904.
60 Hirsch-Behnam A, Delius H, de Villiers EM: A comparative sequence analysis of two human papillomavirus (HPV) types 2a and 57. Virus Res 1990;18:81–98.
61 Chong T, Apt D, Gloss B, Isa M, Bernard HU: The enhancer of human papillomavirus type 16: Binding sites for the ubiquitous transcription factors oct-1, NFA, TEF-1, NF1, and AP-1 participate in epithelial cell specific transcription. J Virol 1991;65:5933–5943.
62 Tan SH, Gloss B, Bernard HU: During negative regulation of the human papillomavirus-16 E6 promoter, the viral E2 protein can displace Sp1 from a proximal promoter element. Nucleic Acids Res 1992;20:251–256.
63 Lorincz AT, Reid R, Jenson AB, Greenberg MD, Lancaster W, Kurman RJ: Human papillomavirus infection of the cervix: Relative risk associations of 15 common anogenital types. Obstet Gynecol 1992;79:328–337.
64 Pfister H, Hettich I, Runne U, Gissmann L, Chilf GN: Characterization of human papillomavirus type 13 from focal epithelial hyperplasia Heck lesions. J Virol 1983;47:363–366.
65 Beaudenon S, Praetorius F, Kremsdorf D, Lutzner M, Worsaae N, Pehau-Arnauded G, Orth G: A new type of human papillomavirus associated with oral epithelial hyperplasia. J Invest Dermatol 1987;88:130–135.
66 Favre M, Obalek S, Jablonska S, Orth G: Human papillomavirus (HPV) type 50, a type associated with epidermodysplasia verruciformis (EV) and only weakly related to other EV-specific HPVs. J Virol 1989;63:4910.
67 Gissmann L: Human papillomaviruses and genital cancer. Semin Cancer Biol 1992;3:253–261.
68 Pfister H: Human papillomaviruses and skin cancer. Semin Cancer Biol 1992;3:263–271.

Intervirology 1994;37:150–158

Louise T. Chow
Thomas R. Broker

Department of Biochemistry and Molecular Genetics, Schools of Medicine and Dentistry, University of Alabama at Birmingham, Birmingham, Ala., USA

Papillomavirus DNA Replication

Key Words
Human papillomavirus · Bovine papillomavirus · SV40 · E1 protein · E2 protein · Replication protein A · Origin of replication

Biology of Papillomaviruses

Human and animal papillomaviruses have strict species and tissue tropisms. The closed circular genomic DNA is typically 7.6–7.9 kbp. It replicates in squamous epithelium of preferred body sites as multicopy extrachromosomal plasmids and causes benign squamous lesions variously called warts, papillomas or condylomata. Over two dozen human papillomavirus types (HPVs) can infect the anogenital epithelia, constituting the most prevalent sexually transmitted diseases of viral etiology. Lesions caused by HPV6 or 11 rarely progress, whereas infections by high-risk types such as 16, 18 or their close relatives, in relatively exceptional cases, can develop into high-grade dysplasias and carcinomas [1]. In these severe lesions, viruses are no longer produced because viral DNA is often integrated into the host chromosomes, generally accompanied by deletions or mutations, and the tissues do not differentiate to permit late gene expression [2, 3]. As such, HPVs are major human pathogens and it is important to understand their biology so that effective antiviral therapies can be designed.

In situ hybridization of tissue biopsies with riboprobes revealed that abundant viral mRNA transcription and DNA amplification leading to the production of progeny virions are tightly linked to squamous epithelial differentiation regardless of virus type [3–7]. This differentiation dependence is a major factor in the difficulties of studying their life cycle in vitro. Furthermore, few virions are generated in vivo except for viruses such as HPV1 and bovine papillomavirus type 1 (BPV 1) that are trophic for cutaneous epithelia. Only limited gene expression was observed in monolayer cultures of primary keratinocytes infected with HPV 1 [8]. When transfected into proliferating mammalian cells, the cloned HPV DNA is either lost within a few cell divisions or becomes integrated into host chromosomes. Only in rare primary human keratinocytes can HPV DNA persist as episomes for an extended period of time [9]. Cells from clinical specimens that harbor episomal HPV DNA are difficult to maintain as monolayers in culture [10]. Virion production has been achieved in nude mice xenografts of infected human tissues or a cell line harboring extrachromosomal HPV DNA derived from a cervical dysplasia [see ref. 11–13 and references therein]. However these systems are not amenable to genetic manipulation. The methodology for recapitulating stratified and differentiated squamous epithelium in vitro has been developed recently [see ref. 14, 15 and references therein]. Yet, virus production in such differentiated keratinocytes is very limited, as in vivo [16, 17]. BPV 1 which causes fibropapillomas in cattle proves to be an exception in several respects. Although no virus particles are produced in vitro, infection with virions or transfection with cloned DNA can transform rodent fibroblast cell lines such as C127 or NIH 3T3 in which extrachromosomal plasmids are maintained at a stable 100–200 copies per cell [18]. This heterologous BPV 1/mouse cell system made it possible to identify genes responsible for transformation and those required for stable DNA replication.

The patterns of gene expression and DNA replication of human or animal viruses have been examined in productively infected specimens [3–7, 19, 20], in epithelial raft cultures of explanted condylomatous tissues [16], and

L.T. Chow, MD
Department of Biochemistry and Molecular Genetics
Schools of Medicine and Dentistry
University of Alabama at Birmingham
Birmingham, AL 35294-0005 (USA)

in growth-arrested BPV-1-transformed murine cells [21–23], which may resemble, to a degree, the noncycling differentiated keratinocytes. Patterns of BPV 1 DNA replication immediately after transfection into murine cells [24] have also been informative. Each of these approaches provides a view through a narrow window. We propose that there are several modes of viral DNA replication during the virus infectious cycle. (1) Viruses gain entry into the basal stem cells of permissive cutaneous or mucosal squamous epithelium after wounding. Wound healing promotes the proliferation of basal cells and also activates the expression of the viral early genes, leading to a temporary amplification and establishment of the DNA plasmids. (2) After the wound has healed, most basal cells return to a state of quiescence and divide only infrequently to replace the parabasal transit amplifying cells. The stratified epithelium is replenished by cell divisions of the parabasal cells, similar to an uninfected epithelium [25], except that the cell divisions are more frequent. The viral transcriptional activity is autoregulated to a low level and viral DNA replicates only when these basal and parabasal cells enter into the S phase of the cell cycle, resulting in a steady state of low copy numbers of virus DNA, generally thought to be less than 20–50 per cell. (3) After the cells ascend, they leave the cell cycle and undergo progressive differentiation. Viral transcription and DNA replication can, but do not always, increase in the upper spinosum. The progression of the differentiation program of the epithelium becomes retarded, resulting in an increased thickness of the spinous cells to produce warty or condylomatous tissues. Warts may remain for years, but they often regress into a persistent subclinical state due to intervention by the host's immune system. Conversely, latent infections may become reactivated following local or general immunosuppression or during wound healing. It is the controlled mode of autonomous DNA replication in the basal and parabasal epithelial cells in an established wart or in the subclinical state that has been considered a model for studying the regulation of host cell DNA replication.

The article summarizes work from many laboratories concerning the identification of viral and host replication proteins, the properties of the viral replication proteins, the sequence arrangement in the origin of replication *(ori),* and the consequence of mutations in cis elements or trans factors of both bovine and human papillomaviruses. These investigations include genetic analysis of BPV 1 replication in transformed murine cells, transient replication of BPV 1 and HPVs in transfected cells, and cell-free replication of BPV 1 and HPV 11 with heterologous and homologous cell extracts, respectively. A comparison is also made to the replication of SV40 which has been well characterized.

Papillomaviral Proteins Required for Episomal DNA Replication

Similar to SV40, papillomaviruses contribute to the replication complex only those proteins that recognize the *ori* and promote initiation. All other replication enzymes and proteins are supplied by the host cell. Unlike the multifunctional SV40 T-antigen (T-Ag), which initiates viral DNA replication, interacts with tumor suppressor proteins p53 and pRB leading to cellular transformation, and regulates viral promoters [for a review see ref. 26], papillomaviruses segregate these functions into four proteins, E1, E2, E6, and E7. E1 and E2 have been identified as the origin-recognition replication proteins in two replication assays, consistent with previous genetic analyses [see ref. 27–29 and references therein]. One assay is transient replication in transfected cells following efficient introduction by electroporation of an *ori* reporter plasmid and strong E1 and E2 expression vectors [30–33]. The other assay is cell-free replication using mammalian cell extracts supplemented with purified E1 and E2 proteins expressed in insect Sf9 cells from recombinant baculoviruses [34, 35]. The segregation of functions into multiple papillomavirus proteins makes it possible to dissect the functional domains of each protein and provides an excellent opportunity to examine the complexities and the regulation of the initiation of viral DNA replication as well as the molecular basis of viral pathogenesis.

The Family of E2 Proteins

All papillomaviruses encode two or more E2 proteins as a result of alternative promoter usage and mRNA splicing. Each has long been recognized for its role in transcriptional regulation [ref. 36, and for reviews see ref. 37, 38]. Only the full-length E2 protein can support replication [30, 34, 39, 40]. The full-length E2 protein of human or animal papillomaviruses migrates as a 43- to 48-kD species in SDS-PAGE. It consists of three functional domains, a trans-acting domain in the amino-terminal portion approximately 200 amino acids long, a central flexible hinge which is highly divergent in length and sequence among papillomaviruses and overlaps the E4 gene in a different reading frame, and a DNA-binding and protein dimerization domain located in the 90 carboxyl-terminal residues [41, 42].

The E2 protein binds as a dimer to a palindromic consensus sequence $ACCN_6GGT$, alternatively termed the E2-binding site (BS) or E2-responsive sequence (RS) [42–45]. The affinity for E2BS in vitro varies with the specific sequence and is generally stronger when the motif is $ACCGN_4CGGT$ and the central bases are A/T. Weak binding to $ACCN_6GTT$ has also been reported for BPV 1 [46]. There are multiple copies of the E2BS in all known papillomaviruses and, as discussed below, at least some are critical to the activity and regulation of the viral *ori*.

The BPV 1 E2 protein is a phosphoprotein. A mutation at one of the phosphorylation sites (Ser301), which is not generally conserved among the papillomaviruses, increases the plasmid copy number in transformed cells [47]. However, it is not known whether the mutation affects replication directly. Although most mutations that eliminate the ability of the E2 protein to bind DNA are unable to support transient *ori* replication [40, 48], some E2 C-terminal truncation mutations do, albeit with a reduced efficiency [40]. The discrepancy has not been resolved. Several factors may be responsible: the high affinity with which BPV 1 E1 protein independently associates with the *ori* (see below), the amount and stability of the E1 and the mutated E2 proteins in the transfected cells, and the ability of the E1 protein to interact properly with the particular mutated E2 protein (see below).

Other forms of E2 protein that contain the DNA-binding domain but different N-terminal domains are transcription repressors [36, 42, 45, 49–55], and some have been shown to repress transient replication as well [39]. Repression might be due to competitive binding to the E2BS or to heterodimer formation with the full-length E2 protein as shown for transcriptional regulation [55, 56]. However cell-free replication indicated that BPV-1 E2:E2C heterodimer was able to support replication [D. Lim, M. Gossen, R. Mendoza, M. Botchan, pers. commun.]. A BPV-1 mutation unable to encode the E2-Tr protein (an N-terminal truncated E2 protein) replicates to high copy number in transformed cells [57, 58]. But this result may be a combination of a direct alleviation of repression of DNA replication and an indirect effect due to elevated transcription of replication genes.

The E1 Protein

The E1 proteins of papillomaviruses share moderate to high homology among 19 papillomavirus sequences that were aligned and are typically between 600 and 650 amino acids long [Q.L. Zhu, T.R. Broker et al., in prep.] but they have variable mobility in SDS-PAGE, with an apparent molecular weight ranging from 68 to 85 kD. The biochemical properties of the BPV 1 E1 protein have been examined in detail in several reports using proteins purified from insect Sf9 cells infected with recombinant baculoviruses, from *Escherichia coli* as a fusion protein, or by using in-vitro-translated proteins. In essence, the BPV 1 E1 protein has many of the properties of the SV40 T-Ag in its role as a replication initiation protein [for a review see ref. 59]. It binds specifically to the *ori* and also to DNA nonspecifically [34, 60–65]. It is a DNA-dependent ATPase and an ATP-dependent helicase [66, 67]. The SV40 T-Ag interacts with the host cell DNA polymerase α/primase [see ref. 68–70, and references therein]. The BPV 1 E1 protein also specifically contacts this host replication enzyme complex to initiate replication [71]. The BPV 1 E1 protein is a phosphoprotein [72, 73]. Because the state of phosphorylation of SV40 T-Ag at specific residues regulates its ability to function in DNA replication [26], it is conceivable that similar regulation also exists for the E1 protein. But mutation at the $p34^{cdc2}$ phosphorylation site did not affect replication [74]. Whether additional phosphorylation sites regulate activity remains to be investigated.

Much less is known about the biochemical properties of the HPV E1 protein. Given the high degree of homology of E1 proteins among all the papillomaviruses, it is expected that HPV E1 protein will have similar activities. Immunoprecipitated HPV 11 E1 protein expressed from recombinant baculovirus indeed exhibited ATPase activity [75], and a bacterially expressed HPV 6 E1-maltose fusion protein possesses helicase and ATPase activities [76] but binding to the homologous HPV *ori* has not been consistently reported [75, 77]. Replication assays suggest that binding to the *ori* is at best very weak (to be discussed).

The 5′ portion of the BPV 1 E1 open reading frame (ORF) also encodes a 23-kD modulator (M) protein from a spliced message which terminates translation shortly downstream of the splice acceptor site [78]. Both the splice donor and acceptor sites are highly conserved among papillomaviruses, and the corresponding splice has been identified in a transcript from an HPV 11 condyloma [50]. A splice donor site mutation unable to produce this protein did not affect stable BPV 1 plasmid replication in transformed cells [79]. HPV 11 encodes an E1M^E2C protein consisting of the amino-terminal portion of the E1 ORF joined to the DNA-binding domain of the E2 protein. It represses transcription and replication in transfected cells [36, 39]. Interestingly, the E1M domain corresponds precisely to the BPV-1 M protein. The in vivo function of either of these alternative E1 proteins is not known.

```
         (HPV E2BS#2)                      **        E1BS               **                 (HPV E2BS#3        E2BS#4)

         .43                                 1..7945                                                          7880.
BPV-1:   TTTTCCCGCTTGAAAAAACGGTGATGGTG   TGATTATTGTTAACAACAATTATTCACTGGGAAAAAATACATAGTCTTTACTTACCGGTTTCGGTGAGCTTAAAATATAA

         .47                                    1..7868                                                       7812.
CRPV:    ACCGATATCGGTTTCTGG CCATTACATG   TTTCTTAATTATTGTTAGCAACAACCATCA  TTAAATCAAAGTGCA TAACACCGCATGCGGTGTGAGTCAATACTTA

         .7774                         7815..1                                                                  64.
HPV-1:   ACCGAATTCGGTCGTAAAAATCTAAAGTGA TGATT  GTTGTTGTTAACTACCATCATTC  ATTATTCTAGTTACAACAAGAACCTAGGAGTTATATGCCAGAAGTAA

         .7986                         8027..1                                                                  61.
HPV-7:   ACCGCTTTCGGTTGCT ACACGTTTTATTA  CTTTCTAGTTATTATGTTTAATAATTGTACAGTTACTAAA      GGGCGTAACCGAAAACGGTCCG ACCGAAAACGGT

         .7797                                  7851..1                                                         50.
HPV-35:  ACCGATTTGGATTACAGTTTTATAAG  CATTTCTTTTTATTATAGTTAGTAACAATCCC    TATAAAAAAAACAGGGAGTGACCGAAAACGGTC GTACCGAAAACGGT

         .7868                           7912..1                                                                63.
HPV-31:  ACCGTTTTCGGTTACAGTTTTACAAG  CAATTGTTCTTTTTATACTTAATAATAATAATCTTAGTATAAAAAAGTAGGGAGTGACCGAAAGTGGT  GAACCGAAAACGGT

         .7859                             7902..1                                                              61.
HPV-16:  ACCGTTTTGGGTTACACATTTACAAG  CAACTTATATAATAATACTAAACTACAATAATTCATGTA TAAAACTAAGGGCGTAACCGAAATCGGTT GAACCGAAACCGGT

         .7780                        7824..1                                                                   66.
HPV-58:  ACCGTTTTGGGTCACATTGTTCATGTTTCAACATTTTATATAATACTAAACTATAATGCCAAATCTTGTAAAAACTAGGGTGTAACCGAAAACGGTCTG ACCGAAACCGGT

         .7866                         7909..1                                                                  65.
HPV-33:  ACCGTTTTAGGTCATATTGGTCATTTA TAATCTTTTATATAATAGTAAACTATAATGCCA AGTTTTAAAAAAGTAGGGTGTAACCGAAAGCGGTTC AACCGAAAACGGT

         .7760                                 7808..1                                                          58.
HPV-51:  ACCGATTTGTGTTCTGCCTATGCTTG  CAACATTTTTCTTATAACATTTAACAATTATC  TTGTAAAAACTA  GGGTGTAACCGAAAAGGGTTATGACCGAAAACGGT

         .7798                      7833. .1                                                                    70.
HPV-39:  ACCGAAATAGGTTGGGCAT ACATACC TATACTTTTA CTTATAACATTTTATAAGTATCTTGTTTAAAAAAA   GGGAGTAACCGAAAACGGTCAGGACCGAAATCGGT

         .7822                      7857. .1                                                                    69.
HPV-18:  ACCGAAATAGGTTGGGCAGCACATAC  TATACTTTTC ATTAATACTTTTAACAATTGT  AGTATATAAAAAA  GGGAGTAACCGAAAACGGTCGGGACCGAAAACGGT

         .7820                        7758..1                                                                   66.
HPV-45:  ACCGAAAAAGGTTGGGCCCTATAA  CACATACCTTTT CTTAATACTTTTAACAATTATAC  TACATAAAAAA  GGGTGTAACCGAAAACGGTTCGAACCAAAAACGGT

         .7861                         7902..1                                                                  61.
HPV-6:   ACCGGTTTCGGTTAT CCACA  CCCTACATATTTCCTTCTTATAGTTAATAACAATCTT  GGTTTAAAAAATA  GGAGGG ACCGAAAACGGTTC AACCGAAAACGGT

         .7892                         7933..1                                                                  61.
HPV-11:  ACCGGTTTCGGTTAC CCACA  CCCTACATATTTCCTTCTTATACTTAATAACAATCTT  AGTTTAAAAAAGA  GGAGGG ACCGAAAACGGTTC AACCGAAAACGGT
```

Fig. 1. Sequence alignment of the *ori* of human and animal papillomaviruses. The E2BS is underlined and overlined. The BPV-1 E1BS and putative HPV E1BS are in bold and double underlined. Dots are over nucleotides for which the genomic positions are given. Note that the BPV1 and CRPV sequences are in reverse orientation relative to those of the HPVs. ** Denotes AT-rich regions that are located on one or both sides of the E1BS. CRPV = Cottontail rabbit papillomavirus.

The *ori*

The minimal BPV 1 *ori* contains an E1BS which is an 18-bp imperfect palindrome [see ref. 65, and references therein], an AT-rich region, and half of a low-affinity E2BS as determined by both transient and cell-free replication assays [30, 34]. A minimal HPV 11 *ori* has been localized to a comparable region of the viral DNA by transient replication [39] and overlaps the E6 promoter [54]. This HPV 11 *ori* also contains an AT-rich region and a putative E1BS, which has a 4-bp mismatch to the BPV 1 E1BS, flanked by several copies of the E2BS. In fact, this sequence arrangement is highly conserved among all papillomaviruses (fig. 1). However, unlike the BPV 1 *ori,* HPV 11 *ori* activity requires at least one intact E2BS. A strong *ori* was rendered totally inactive when half-sites of all three native E2BSs (E2BS No. 2, 3, and 4 in fig. 1) were mutated, when tested in either replication assay [35, 39]. Furthermore, the efficiency of transient replication appears to be proportional to the number of copies of E2BS in the template [32, 39]. Subsequent studies confirmed that one or more E2BSs are required in transient HPV *ori* replication [77, 80]. Surprisingly, a single copy of a synthetic E2BS can initiate a low level of replication in either assay [J.S. Liu, S.R. Kuo, T.R. Broker, L.T. Chow, unpubl. results]. A recent investigation also demonstrated that the ability of mutated BPV 1 E2 protein or an altered E2BS to support transient replication is proportional to the affinity between them [48]. A low-affinity E2BS can function only a short distance from the E1BS whereas, with a longer separation, only a high-affinity E2BS can support replication, consistent with an interaction between the two viral proteins, as discussed below.

Interactions among E1, E2 Proteins and the *ori*

The interactions among BPV 1 E1 protein, E2 protein and the *ori* and their roles in replication have been intensively studied using proteins from different sources, as described above. The results can be summarized as follows. (1) The BPV 1 E1 protein may bind as an oligomer in that oligomer binding exhibits cooperativity [64]. Binding to the *ori* is essential for replication, as insertion mutations at the E1BS eliminate both binding and replication [30, 34], with few exceptions [81]. (2) The E1 protein is an ATPase and helicase. In the presence of ATP, $MgCl_2$, topoisomerase I and the single-stranded DNA-binding protein RPA (i.e., SSB), it converts the form I DNA into an unwound U form [66, 67]. These properties are virtually identical to those of SV40 T-Ag [82], which binds to the SV40 *ori* as a double hexamer [83]. It is in this topologically unwound template that the first labeled nascent strand is found in cell-free SV40 replication [84]. (3) The E2 protein is an essential component of the initiation process. It interacts with E1, stabilizes its interaction with E1BS and stimulates *ori* unwinding [34, 62–64], tethers RPA [85], and eliminates the nucleosome-mediated repression of replication [86]. In the case of HPV 11 replication, the E2 protein, rather than the E1 protein, is the primary *ori* recognition protein [35, 39]. Thus, these properties of the full-length E2 proteins distinguish them from transcription factors that increase the efficiency of replication of SV40 DNA [for a review see ref. 87]. These major conclusions are further elaborated below.

BPV 1 or HPV E1 and E2 proteins expressed in insect Sf9 cells interact to form heteromeric complexes, as detected by coimmunoprecipitation or the McKay assay [73, 77, 88–90]. It was initially postulated that the 68-kD BPV 1 E1 protein was targeted to the BPV 1 *ori* via the full-length BPV 1 E2 protein bound to the E2BS [88]. This view was modified to favor a primary role of the E1 protein in *ori* recognition when the BPV 1 E1 protein was found to bind specifically to an 18-bp imperfect palindrome spanning nucleotide position 1 of BPV 1 DNA and when it was observed that the minimal BPV 1 *ori* had only one-half of a low-affinity E2BS but an intact E1BS was essential. These unexpected observations were interpreted to reflect the formation of an E1/E2 complex on the E1BS, which is stabilized by association of the E2 protein with the nearby half E2BS. However, as discussed, subsequent studies of both HPV and BPV 1 replication with either assay demonstrate that the E2 binding to an intact E2BS is very important indeed, more so for HPV than for BPV 1.

The BPV 1 E1 protein binding to the *ori* is greatly stimulated and stabilized by the addition of BPV 1 E2 protein. The interaction is cold-sensitive and is affected by the distance between the two protein-binding sites [63, 64]. Further, in the absence of an E2BS, addition of the BPV 1 E2 protein inhibited BPV 1 E1 binding to the origin, suggesting that E1-E2 complexes formed in the absence of DNA-binding sites may not be the same as those that form on the *ori*. This conclusion is consistent with the observation that the E1 protein domain required for E2 stimulation [63] does not coincide with the E1-E2 interaction domain identified previously by coimmunoprecipitation [90]. This E1 protein domain responsive to E2 stimulation co-localizes with the specific DNA-binding domain between residues 180 and 299. E1 deletion mutations that do not bind to the *ori* in the presence or absence of E2 protein do not support replication in transient assays, and certain substitution mutations in amino acids that are highly conserved among papillomaviruses in this region affect both functions [63].

The domain in the E2 protein which interacts with the E1 protein is thought to be the N-terminal portion, because the alternative forms of E2 lacking that domain do not support *ori* replication [30, 31, 34]. A BPV 1 E2 protein deletion of the hinge region is unable to support *ori* replication [40], although this region did not seem to play a critical role in transcription regulation [40, 41, 55]. Despite the highly divergent hinge region among papillomaviruses, mixed combinations of HPV 11 and BPV 1 proteins can replicate all papillomavirus *ori*s tested in the transient assay [31]. We suspect that flexibility afforded by the hinge domain, rather than specific amino acid residues in the hinge, is important to facilitate proper interactions with the *ori*, the viral E1 protein or host replication proteins.

Similar mutagenic studies on the HPV E1 and E2 proteins have not been performed, nor has detailed biochemical characterization of the HPV E1 protein. But the facts that an E2BS is essential and that insertion or deletion mutations of the putative E1BS reduced but did not eliminate *ori* replication, whereas similar mutations in the BPV 1 *ori* almost always abolish replication [35, 39, 77, 80; Liu et al., unpubl. results] indicate that the HPV E1 protein alone does not bind to the *ori* with an affinity approaching that of BPV 1 E1 protein for its recognition sequence. This difference helps explain the ability of BPV 1 E1 protein to repress the BPV 1 E6 promoter [91–93], whereas the HPV 11 E1 is not a repressor of the homologous viral E6 promoter [94]. Irrespective of these differences between HPV and BPV, the studies demonstrate that both viral proteins and one or more copies of their binding sites are crucial for *ori* replication.

As with SV40 replication [95, 96], papillomavirus replication is bidirectional with θ form intermediates in transformed cells and in cell-free replication as assessed by one- and two-dimensional gel electrophoresis [34, 97, 98, 98a]. Once DNA replication is initiated, the E1 protein-DNA interaction probably changes from a sequence-specific DNA-binding protein to that of a sequence-nonspecific binding protein and the E1 protein may remain associated with the replication fork and unwind the DNA ahead of the diverging replication forks [Kuo et al., unpubl. results], by analogy to SV40 replication [70].

Cell-Free Replication

Cell-free BPV 1 *ori* replication has been established with mouse FM3A or HeLa cell extracts supplemented with BPV 1 E1 and E2 proteins purified from Sf9 cells infected by recombinant baculoviruses [34, 85, 86]. The E1 protein was tagged at the amino-terminus with an epitope rich in glutamic acid derived from the polyoma middle T-Ag (i.e., the EE epitope) which facilitates protein purification. At low concentrations of the E1 protein alone, the BPV 1 *ori*⁺ template replicates with a low efficiency which is stimulated by the addition of the E2 protein. At high concentrations of E1 protein alone, an *ori*⁻ template also replicates, which is, however, suppressed by addition of the E2 protein. These results can be interpreted in terms of the ability of the E1 protein to bind to and unwind the *ori* while the E2 protein stimulates both activities. As already described, the E2 protein plays additional critical roles. It helps recruit RPA [85]. In addition, it and perhaps other transcription factors alleviate nucleosome-mediated repression of replication [86], as has been demonstrated for the role of transcription factors in SV40 replication [99, 100]. This observation helps explain why BPV 1 E2 protein is absolutely essential in transfected cells.

A cell-free replication system for the HPV 11 *ori* has been established using human 293 cell extracts supplemented with HPV 11 E2 protein and the EE-epitope-tagged HPV 11 E1 protein purified from Sf9 cells [35], the first homologous cell-free replication system for a papillomavirus. Both similarities and differences are found when compared to BPV 1 cell-free replication or to transient replication. For instance, E2 protein stimulates HPV 11 *ori* replication in the presence of low concentrations of the E1 protein. At high concentrations, the E1 protein alone can also initiate replication from an *ori*⁻ template but it is repressed by the addition of the E2 protein. The notable difference is that an intact E2BS is absolutely required, as previously demonstrated in transient assays [39]. The activity of a strong HPV 11 *ori* is reduced by 40-fold or more when the half-site of each of the three dyad E2BSs is mutated. In contrast, the efficiency with which the BPV-1 proteins replicate this mutated *ori* is reduced by less than 4-fold. Furthermore, high concentrations of E1 protein alone do not replicate the HPV 11 *ori*⁺ template much more efficiently than the *ori*⁻ template. These results reenforce the notions that, in the absence of the E2 protein, HPV 11 E1 protein binds weakly, if at all, to the putative E1BS, whereas the BPV 1 E1 protein binds to the homologous and perhaps to the heterologous *ori* sequences. We suggest that, for HPV 11 and perhaps HPVs in general, the E2 protein binds to one or more copies of the E2BS, then tethers the E1 protein, which in turn recruits the host DNA proteins in the assembly of the initiation complex on the *ori*.

A difference with transient replication is also noted. The BPV 1 proteins usually replicate homologous and heterologous *ori*s more efficiently than HPV 11 proteins in transfected cells [31]. In contrast, in the cell-free replication, the two homologous proteins/*ori* combinations are much more effective than the heterologous combinations [35]. These observations suggest that the results of transient assays reflect a more efficient expression or a higher stability of the BPV 1 proteins in the transfected cells. Thus, the *ori* sequences and the viral proteins of each papillomavirus have coevolved to achieve optimal DNA-protein and protein-protein interactions for regulated replication in their respective host cell milieu.

The requirements for host replication proteins have been identified using antibodies or inhibitors to host proteins in the cell-free replication systems. Replication of both BPV 1 and HPV 11 is dependent on the host DNA polymerase α/primase, DNA polymerase δ/proliferating cell nuclear antigen, RPA, and topoisomerases I and II [34, 35]. These factors are also needed for SV40 replication [ref. 101 and references therein] and, no doubt, for host genome replication as well.

Comparison with the SV40 Replication System

Both cell-free replication and transient replication in transfected cells are central to understanding the requirements and mechanisms for papillomavirus replication, as they are for SV40 replication. Mechanistically, replication of these papovaviruses appears to be identical. The E1 protein probably twists and melts the *ori* as the SV40 T-Ag does on the SV40 *ori* [59]. These events are expected because BPV 1 E1 protein also unwinds form I DNA into an unwound U form in the presence of topoisomerase I and single-stranded DNA-binding protein RFA [66, 67]. The difference resides in the specificities in host-viral protein interactions. SV40 DNA replication can only be supported by monkey or human cells, but not by murine cells, a restriction attributed to the inability of the SV40 T-Ag to interact with the mouse RPA [70, 102]. In contrast, the BPV 1 E2 protein can interact with the human RPA [85].

In fact, PV replication in vitro is promiscuous. HPV and BPV 1 *ori*s can replicate transiently in a number of mammalian cell lines including human, monkey, murine and hamster in response to mixed or matched E1 and E2 proteins supplied exogenously from expression vectors [31]. Both murine and human cell extracts can replicate the BPV 1 *ori* in cell-free systems [34, 35, 85]. These results suggest that the strict species and tissue specificity observed in vivo is probably a reflection of limitations on virus entry or viral early gene expression rather than on DNA replication per se.

In summary, the DNA replication of papillomaviruses has been studied in clinical specimens, in transformed cells, following transient transfections, and in cell-free systems. Together, these results suggest that papillomaviruses have three modes of DNA replication: (1) an establishment phase of modest amplification upon entry into basal stem cells of a wounded squamous epithelium; (2) a steady-state, low-level maintenance mode, enough to keep up with division of the basal and parabasal cells, and (3) a vegetative amplification mode for the production of progeny virus in differentiated cells. Both the viral E1 and E2 proteins are required to mark the *ori* and initiate viral DNA replication, while the host cells provide all the other replication enzymes. The *ori,* located in the upstream regulatory region, consists of binding sites for the E1 and E2 proteins. For BPV 1, E1 is the primary *ori* recognition protein whereas for HPV 11, and perhaps for HPVs in general, the E2 protein is the critical *ori* protein.

The papillomaviruses provide a superb system to study eukaryotic DNA replication. Being important human pathogens, they are the subject of intense scrutiny in the search for antiviral pharmaceutical products and further serve as excellent indicators for antitumor drugs for chemotherapy of cancers, many of which are caused by the disregulation of the viral E6 and E7 oncogenes, the products of which interact with and inactivate the host tumor suppressor proteins.

Acknowledgments

The authors are supported by grants from the US Public Health Service CA 36200, CA 43629, AI 34674 and The Council for Tobacco Research-USA 2550. We thank our students, fellows, and collaborators for their many contributions.

References

1 de Villiers EM: Heterogeneity of the human papillomavirus group. J Virol 1989;63:4898–4903.
2 Cullen AP, Reid R, Campion M, Lörincz AT: Analysis of the physical state of different human papillomavirus DNAs in intraepithelial and invasive cervical neoplasm. J Virol 1991; 65:606–612.
3 Stoler MH, Rhodes CR, Whitbeck A, Wolinsky SM, Chow LT, Broker TR: Gene expression of human papillomavirus type 16 and 18 in cervical neoplasias. Hum Pathol 1992;23:117–128.
4 Broker TR, Chow LT, Chin MT, Rhodes CR, Wolinsky SM, Whitbeck A, Stoler MH: A molecular portrait of human papillomavirus carcinogenesis. Cancer Cells 1989;7:197–208.
5 Stoler MH, Whitbeck A, Wolinsky SM, Broker TR, Chow LT, Howett MK, Kreider JW: Infectious cycle of human papillomavirus type 11 in human foreskin xenografts in nude mice. J Virol 1990;64:3310–3318.
6 Stoler MH, Wolinsky SM, Whitbeck A, Broker TR, Chow LT: Differentiation-linked human papillomavirus types 6 and 11 transcription in genital condylomata revealed by in situ hybridization with message-specific RNA probes. Virology 1989;172:331–340.
7 Beyer-Finkler E, Stoler MH, Girardi F, Pfister H: Cell differentiation-related gene expression of human papillomavirus 33. Med Microbiol Immunol 1990;179:185–192.
8 Chow LT, Reilly SS, Broker TR, Taichman LB: Identification and mapping of human papillomavirus type 1 RNA transcripts recovered from plantar warts and infected epithelial cell cultures. J Virol 1987;61:1913–1918.
9 Mungal S, Steinberg BM, Taichman LB: Replication of plasmid-derived human papillomavirus type 11 DNA in cultured keratinocytes. J Virol 1992;66:3220–3224.
10 DiLorenzo TP, Taichman LB, Steinberg BM: Replication and persistence of HPV DNA in cultured cells derived from laryngeal papillomas. Virology 1992;186:148–153.
11 Kreider JW, Howett MK, Leure-Dupree AE, Zaino RJ, Weber JA: Laboratory production in vivo of infectious human papillomavirus type 11. J Virol 1987;61:590–593.
12 Kreider JW, Patrick SD, Cladel NM, Welsh PA: Experimental infection with human papillomavirus type 1 of human hand and foot skin. Virology 1990;177:415–417.
13 Sterling J, Stanley M, Gatward G, Minson T: Production of human papillomavirus type 16 virions in a keratinocyte cell line. J Virol 1990; 64:6305–6307.
14 Kopan R, Traska G, Fuchs E: Retinoids as important regulators of terminal differentiation: Examining keratin expression in individual epidermal cells at various stages of keratinization. J Cell Biol 1987;105:427–440.

15 Wilson JL, Dollard SC, Chow LT, Broker TR: Epithelial-specific gene expression during differentiation of stratified primary human keratinocyte cultures. Cell Growth Differ 1992;3: 471–483.
16 Dollard SC, Wilson JL, Demeter LM, Bonnez W, Reichman RC, Broker TR, Chow LT: Production of human papillomavirus and modulation of the infectious program in epithelial raft cultures. Genes Dev 1992;6:1131–1142.
17 Meyers C, Frattini MG, Hudson JB, Laimins LA: Biosynthesis of human papillomavirus from a continuous cell line upon epithelial differentiation. Science 1992;257:971–973.
18 Law MF, Lowy DR, Dvoretzky I, Howley PM: Mouse cells transformed by bovine papillomavirus contain only extrachromosomal viral DNA sequences. Proc Natl Acad Sci USA 1981;78:2727–2731.
19 Spalholz BA, Howley PM: Papillomavirus-host cell interactions. Adv Viral Oncol 1989;8:27–53.
20 Zeltner R, Borenstein LA, Wettstein FO, Iftner T: Changes in RNA expression pattern during the malignant progression of cottontail rabbit papillomavirus-induced tumors in rabbits. J Virol 1994;68:3620–3630.
21 Roberts JM, Weintraub H: Cis-acting negative control of DNA replication in eukaryotic cells. Cell 1988;52:397–404.
22 Burnett S, Kiessling U, Pettersson U: Loss of bovine papillomavirus DNA replication control in growth-arrested transformed cells. J Virol 1988;63:2215–2225.
23 Burnett S, Strom AC, Jareborg N, Alderborn A, Dillner J, Moreno-Lopez J, Pettersson U, Kiessling U: Induction of bovine papillomavirus E2 gene expression and early region transcription by cell growth arrest, correlation with viral DNA amplification and evidence for differential promoter induction. J Virol 1988;64: 5529–5541.
24 Berg L, Lusky M, Stenlund A, Botchan MR: Repression of bovine papillomavirus replication is mediated by a virally encoded trans-acting factor. Cell 1986;46:753–762.
25 Demeter LM, Stoler MH, Broker TR, Chow LT: Induction of proliferating cell nuclear antigen (PCNA) in differentiated keratinocytes of human papillomavirus-infected lesions. Hum Pathol 1994;25:343–348.
26 Fanning E: Simian virus 40 large T antigen: The puzzle, the pieces, and the emerging picture. J Virol 1992;66:1289–1293.
27 Sarver N, Rabson MS, Yang YC, Byrne JC, Howley PM: Localization and analysis of bovine papillomavirus type 1 transforming function. J Virol 1984;52:377–388.
28 Rabson MS, Yee C, Yang YC, Howley PM: Bovine papillomavirus type 1 3′ early region transformation and plasmid maintenance functions. J Virol 1986;60:626–634.
29 DiMaio D, Settleman J: Bovine papillomavirus mutant temperature sensitive for transformation, replication and transactivation. EMBO J 1988;7:1197–1204.
30 Ustav M, Stenlund A: Transient replication of BPV-1 requires two viral polypeptides encoded by the E1 and E2 open reading frames. EMBO J 1991;10:449–457.
31 Chiang CM, Ustav M, Stenlund A, Ho TF, Broker TR, Chow LT: Viral E1 and E2 proteins support replication of homologous and heterologous papillomaviral origins. Proc Natl Acad Sci USA 1992;89:5799–5803.
32 Remm M, Brain R, Jenkins JR: The E2 binding sites determine the efficiency of replication for the origin of human papillomavirus type 18. Nucleic Acids Res 1992;20:6015–6021.
33 Del Vecchio A, Romanczuk H, Howley PM, Baker CC: Transient replication of human papillomavirus DNAs. J Virol 1992;66:5949–5958.
34 Yang L, Li R, Mohr IJ, Clark R, Botchan M: Activation of BPV-1 replication in vitro by the transcription factor E2. Nature 1991;353:628–632.
35 Kuo SR, Liu JS, Broker TR, Chow LT: Cell-free replication of the human papillomavirus DNA with homologous viral E1 and E2 proteins and human cell extracts. J Biol Chem 1994;269:24058–24065.
36 Chiang CM, Broker TR, Chow LT: An E1M^E2C fusion protein encoded by human papillomavirus type 11 is a sequence-specific transcription repressor. J Virol 1991;65:3317–3329.
37 Ham J, Dostatni N, Gauthier JM, Yaniv M: The papillomavirus E2 protein: A factor with many talents. Trends Biochem Sci 1991;16: 440–444.
38 McBride AA, Romanczuk H, Howley PM: The papillomavirus E2 regulatory proteins. J Biol Chem 1991;266:18411–18414.
39 Chiang CM, Dong G, Broker TR, Chow LT: Control of human papillomavirus type 11 origin of replication by the E2 family of transcription regulatory proteins. J Virol 1992;66:5224–5231.
40 Winokur PL, McBride AA: Separation of the transcriptional activation and replication functions of the bovine papillomavirus-1 E2 protein. EMBO J 1992;11:4111–4118.
41 Giri I, Yaniv M: Structural and mutational analysis of E2 trans-activating proteins of papillomaviruses reveals three distinct functional domains. EMBO J 1988;7:2823–2829.
42 McBride AA, Schlegel R, Howley PM: The carboxy-terminal domain shared by the bovine papillomavirus E2 transactivator and repressor proteins contains a specific DNA binding activity. EMBO J 1988;7:533–539.
43 Androphy EJ, Lowy DR, Schiller JT: Bovine papillomavirus E2 trans-acting gene product binds to specific sites in papillomavirus DNA. Nature 1987;325:70–73.
44 Hirochika H, Hirochika R, Broker TR, Chow LT: Functional mapping of the human papillomavirus transcriptional enhancer and its complexes with the trans-acting E2 protein. Genes Dev 1988;2:54–67.
45 Hubbert NL, Schiller JT, Lowy DR, Androphy DJ: Bovine papillomavirus-transformed cells contain multiple E2 proteins. Proc Natl Acad Sci USA 1989;85:5864–5868.
46 Li R, Knight JD, Bream G, Stenlund A, Botchan MR: Specific recognition nucleotides and their DNA context determine the affinity of E2 protein for 17 binding sites in the BPV-1 genome. Genes Dev 1989;3:510–526.
47 McBride AA, Howley PM: Bovine papillomavirus with a mutation in the E2 serine 301 phosphorylation site replicates at a high copy number. J Virol 1991;65:6528–6534.
48 Ustav E, Ustav M, Szymanski P, Stenlund A: The bovine papillomavirus origin of replication requires a binding site for the E2 transcriptional activator. Proc Natl Acad Sci USA 1993; 90:898–902.
49 Chin MT, Hirochika R, Hirochika H, Broker TR, Chow LT: Regulation of the human papillomavirus type 11 enhancer and E6 promoter by activating and repressing proteins from the E2 open reading frame: Functional and biochemical studies. J Virol 1988;62:2994–3002.
50 Rotenberg MO, Chow LT, Broker TR: Characterization of rare human papillomavirus type 11 mRNAs coding for regulatory and structural proteins using the polymerase chain reaction. Virology 1989;172:489–497.
51 Lambert PF, Hubbert NL, Howley PM, Schiller JT: Genetic assignment of multiple E2 gene products in bovine papillomavirus-transformed cells. J Virol 1989;63:3151–3154.
52 Choe J, Vaillancourt P, Stenlund A, Botchan M: Bovine papillomavirus type 1 encodes two forms of a transcriptional repressor: Structural and functional analysis of new viral cDNAs. J Virol 1989;63:1743–1755.
53 Thierry F, Howley PM: Functional analysis of E2-mediated repression of the HPV18 P_{105} promoter. New Biol 1991;3:90–100.
54 Dong G, Broker TR, Chow LT: Human papillomavirus type 11 E2 proteins repress the homologous E6 promoter by interfering with the binding of host transcription factors to adjacent elements. J Virol 1994;68:1115–1127.
55 McBride AA, Byrne JC, Howley PM: E2 polypeptides encoded by bovine papillomavirus type 1 forms dimers through the common carboxyl-terminal domain: transactivation is mediated by the conserved amino-terminal domain. Proc Natl Acad Sci USA 1989;86:510–514.
56 Barsoum J, Prakash SS, Han P, Androphy EJ: Mechanism of action of the papillomavirus E2 repressor: Repression in the absence of DNA binding. J Virol 1992;66:3941–3945.
57 Lambert PF, Monk BC, Howley PM: Phenotypic analysis of bovine papillomavirus type 1 E2 repressor mutants. J Virol 1990;64:950–956.
58 Riese DJ II, Settleman J, Neary K, DiMaio D: Bovine papillomavirus E2 repressor mutant displays a high-copy-number phenotype and enhanced transforming activity. J Virol 1990; 64:944–949.
59 Borowiec JA, Dean FB, Bullock PA, Hurwitz J: Binding and unwinding – How T antigen engages the SV40 origin of DNA replication. Cell 1990;60:181–184.

60 Ustav M, Ustav E, Szymanski P, Stenlund A: Identification of the origin of replication of bovine papillomavirus and characterization of the viral origin recognition factor E1. EMBO J 1991;10:4321–4329.
61 Wilson V, Ludes-Meyers J: A bovine papillomavirus E1-related protein binding specifically to bovine papillomavirus DNA. J Virol 1991; 65:5314–5322.
62 Seo YS, Muller F, Lusky M, Gibbs E, Kim HY, Phillips B, Hurwitz J: Bovine papilloma virus (BPV)-encoded E2 protein enhances binding of E1 protein to the BPV replication origin. Proc Natl Acad Sci USA 1993;90:2865–2869.
63 Thorner LK, Lim DA, Botchan MR: DNA-binding domain of bovine papillomavirus type 1 E1 helicase: Structural and functional aspects. J Virol 1993;67:6000–6014.
64 Lusky M, Hurwitz J, Seo YS: Cooperative assembly of the bovine papilloma virus E1 and E2 proteins on the replication origin requires an intact E2 binding site. J Biol Chem 1993; 268:15795–15803.
65 Holt SE, Schuller G, Wilson VG: DNA binding specificity of the bovine papillomavirus E1 protein is determined by sequences contained within an 18-base-pair inverted repeat element at the origin of replication. J Virol 1994;68: 1094–1102.
66 Seo YS, Muller F, Lusky M, Hurwitz J: Bovine papilloma virus (BPV)-encoded E1 protein contains multiple activities required for BPV DNA replication. Proc Natl Acad Sci USA 1993:90:702–706.
67 Yang L, Mohr I, Fouts E, Lim DA, Nohaile M, Botchan M: The E1 protein of bovine papilloma virus 1 is an ATP-dependent DNA helicase. Proc Natl Acad Sci USA 1993;90:5086–5090.
68 Dornreiter I, Erdile LF, Gilbert IU, von Winkler D, Kelly TJ, Fanning E: Interaction of DNA polymerase α-primase with cellular replication protein A and SV40 T antigen. EMBO J 1992; 11:769–776.
69 Dornreiter I, Copeland WC, Wang TSF: Initiation of simian virus 40 DNA replication requires the interaction of a specific domain of human DNA polymerase α with large T antigen. Mol Cell Biol 1993;13:809–820.
70 Murakami Y, Hurwitz J: Functional interaction between SV40 T antigen and other replication proteins at the replication fork. J Biol Chem 1993;268:11008–11017.
71 Park P, Copeland W, Yang L, Wang T, Botchan MR, Mohr IJ: The cellular DNA polymerase α-primase is required for papillomavirus DNA replication and associates with the viral E1 helicase. Proc Natl Acad Sci USA 1994;91:8700–8704.
72 Sun S, Thorner L, Lentz M, MacPherson P, Botchan M: Identification of a 68-kilodalton nuclear ATP-binding phosphoprotein encoded by bovine papillomavirus type 1. J Virol 1990; 64:5093–5105.
73 Blitz IL, Laimins LA: The 68-kilodalton E1 protein of bovine papillomavirus is a DNA binding phosphoprotein which associates with the E2 transcriptional activator in vitro. J Virol 1991;65:649–656.
74 Lentz MR, Pak D, Mohr I, Botchan MR: The E1 replication protein of bovine papillomavirus type 1 contains an extended nuclear localization signal that includes a $p34^{cdc2}$ phosphorylation site. J Virol 1993;67:1414–1423.
75 Bream GL, Ohmstede CA, Phelps WC: Characterization of human papillomavirus type 11 E1 and E2 proteins expressed in insect cells. J Virol 1993;67:2655–2663.
76 Hughes FJ, Romanos MA: E1 protein of human papillomavirus is a DNA helicase/ATPase. Nucleic Acids Res 1993;21:5817–5823.
77 Lu JZJ, Sun YN, Rose RC, Bonnez W, McCance DJ: Two E2 binding sites (E2BS) alone or one E2BS plus an A/T-rich region are minimal requirements for the replication of the human papillomavirus type 11 origin. J Virol 1993;67:7131–7139.
78 Thorner L, Bucay N, Choe, J Botchan M: The product of the bovine papillomavirus type 1 modulator gene (M) is a phosphoprotein. J Virol 1988;62:2474–2482.
79 Hubert W, Lambert PF: The 23-kilodalton E1 phosphoprotein of bovine papillomavirus type 1 is nonessential for stable plasmid replication in murine C127 cells. J Virol 1993;67: 2932–2937.
80 Sverdrup F, Khan SA: Replication of human papillomavirus (HPV) DNA supported by the HPV type 18 E1 and E2 proteins. J Virol 1994; 68:505–509.
81 Spalholz BA, McBride AA, Sarafi T, Quintero J: Binding of bovine papillomavirus E1 to the origin is not sufficient for DNA replication. Virology 1993;193:201–212.
82 Dean FB, Bullock P, Murakami Y, Wobbe CR, Weissbach L, Hurwitz J: Simian virus 40 (SV40) DNA replication: SV40 large T antigen unwinds DNA containing the SV40 origin of replication. Proc Natl Acad Sci USA 1987;84: 16–20.
83 Dean FB, Borowiec JA, Eki T, Hurwitz J: The simian virus 40 T antigen double hexamer assembles around the DNA at the replication origin. J Biol Chem 1992;267:14129–14137.
84 Bullock PA, Seo YS, Hurwitz J: Initiation of simian virus 40 DNA replication in vitro: Pulse-chase experiments identify the first labeled species as topologically unwound. Proc Natl Acad Sci USA 1989;86:3944–3948.
85 Li R, Botchan MR: The acidic transcriptional activation domains of VP16 and p53 bind the cellular replication protein A and stimulate in vitro BPV-1 DNA replication. Cell 1993;73: 1207–1221.
86 Li R, Botchan MR: Acidic transcription factors alleviate nucleosome-mediated repression of HPV-1 DNA replication. Proc Natl Acad Sci USA 1994;91:7051–7055.
87 DePamphilis ML: Origins of DNA replication that function in eukaryotic cells. Curr Opin Cell Biol 1993;5:434–441.
88 Mohr IJ, Clark R, Sun S, Androphy EJ, MacPherson P, Botchan MR: Targeting the E1 replication protein to the papillomavirus origin of replication by complex formation with the E2 transactivator. Science 1990:250:1694–1699.
89 Yang L, Mohr I, Li R, Nottoli T, Sun S, Botchan M: Transcription factor E2 regulates BPV-1 DNA replication in vitro by direct protein-protein interaction. Cold Spring Harbor Symp 1991;56:335–346.
90 Lusky M, Fontane E: Formation of the complex of bovine papillomavirus E1 and E2 proteins is modulated by E2 phosphorylation and depends upon sequences within the carboxyl terminus of E1. Proc Natl Acad Sci USA 1991;88:6363–6367.
91 Lambert PF, Howley PM: Bovine papillomavirus type 1 E1 replication-defective mutants are altered in their transcriptional regulation. J Virol 1988;62:4009–4015.
92 Schiller JT, Kleiner E, Androphy EJ, Lowy DR, Pfister H: Identification of bovine papillomavirus E1 mutants with increased transforming and transcriptional activity. J Virol 1989;63:1775–1782.
93 Sandler AB, Vande Pol SB, Spalholz BA: Repression of bovine papillomavirus type 1 transcription by the E1 replication protein. J Virol 1993;67:5079–5087.
94 Chiang CM, Broker TR, Chow LT: Properties of bovine papillomavirus E1 mutations. Virology 1992;191:964–967.
95 Schirmbeck R, Deppert W: Structural topography of simian virus 40 DNA replication. J Virol 1991;65:2578–2588.
96 Snapka RM, Permana PA, Maquit G, Shin CG: Two dimensional agarose gel analysis of simian virus 40 DNA replication intermediates. Methods: A Companion to Methods in Enzymology 1991;3:73–82.
97 Schvartzman JB, Adolph S, Martin-Parras L, Schildkraut CL: Evidence that replication initiated at only some of the potential origins in each oligomeric form of bovine papillomavirus type 1 DNA. Mol Cell Biol 1990;6:3078–3086.
98 Yang L, Botchan M: Replication of bovine papillomavirus type 1 DNA initiates within an E2 responsive enhancer element. J Virol 1990;64:5903–5911.
98a Auborn KJ, Little RD, Platt THK, Vaccariello MA, Schildkraut CL: Replicative intermediates of human papillomaviruses type 11 in laryngeal papillomas: Site of replication initiation and direction of replication. Proc Natl Acad Sci USA 1994;91:7340–7344.
99 Cheng L, Kelly TJ: The transcriptional activator nuclear factor I stimulates replication of SV40 minichromosomes in vivo and in vitro. Cell 1989;59:541–551.
100 Cheng LJ, Workman R, Kinston E, Kelly TJ: Regulation of DNA replication in vitro by the transcriptional activation domain of GAL4-VP16. Proc Natl Acad Sci USA 1992;89:589–593.
101 Waga S, Stillman B: Anatomy of a DNA replication fork revealed by reconstitution of SV40 replication in vitro. Science 1994;369: 207–212.
102 Melendy T, Stillman B: An interaction between replication protein A and SV40 T antigen appears essential for primosome assembly during SV40 DNA replication. J Biol Chem 1993;268:3389–3395.

Intervirology 1994;37:159–167

P.G. Fuchs
H. Pfister

Institut für Klinische und Molekulare Virologie der Friedrich-Alexander-Universität Erlangen-Nürnberg, Deutschland

Transcription of Papillomavirus Genomes

Key Words
Enhancer
E2 protein
Promoter
Silencer
Transcription factor
Transcripts

Papillomavirus genomes are transcribed into multiple, differentially spliced and partially overlapping RNA species. The careful analysis of specific virus types usually disclosed several promoters of early and late gene expression. Transcription is subject to complex control by viral and cellular functions and is decisively determined by the differentiation state of the host keratinocyte. As demonstrated for the late genes of bovine papillomavirus (BPV) type 1, promoter activity, transcription termination, RNA processing and turnover can be targets for regulation [1, 2]. In human papillomaviruses (HPVs), studies of transcription control have concentrated thus far primarily on the interaction of promoter sequences with cellular and viral transcription factors.

Promoters

A comprehensive analysis of BPV 1 transcription revealed at least seven promoters [1]. Six are involved in transcription of the early genes; three of them are located in the noncoding region (NCR) of the genome whereas the others appear integrated into early genes. The single late promoter P_L has been mapped to the NCR, close to the 3′ end of the L1 gene. Compared with BPV 1, there is still rather little information on the promoter elements in HPVs. A promoter in front of the open reading frame (ORF) E6 seems to be a common feature of all HPVs studied by mRNA cap site mapping [3–6]. In HPVs 6, 8 and 11, there is an additional promoter within ORF E6 believed to serve the independent expression of ORF E7 [5, 7]. Three intragenic promoters have recently been reported for HPV 18 in front of the E2 gene, within the E2 coding sequences and at the end of the L2 ORF [8]. There are also indications for active promoter elements within the ORF E6 of HPV 1 and 11 [9, 10], as well as in front of ORF E1 of HPV 16 [11].

In cutaneous papillomaviruses of humans and animals, the late promoters are consistently localized within the 5′ part of the NCR [1, 6, 12, 13]. All these promoters present similar sequence features and their transcripts are similarly processed [6]. From the analysis of L1-specific mRNAs of genital HPVs there is some indication for a late promoter in the 5′ part of ORF E1 [11, 14, 15].

Taken together, the presently available data for HPVs allows us to foresee the characterization of further promoter elements, probably leading to a similar plurality as in BPV 1.

P.G. Fuchs
Institut für Klinische und Molekulare Virologie
Friedrich-Alexander-Universität
Schlossgarten 4
D–91054 Erlangen (Germany)

0300–5526/94/0374–0159
$8.00/0

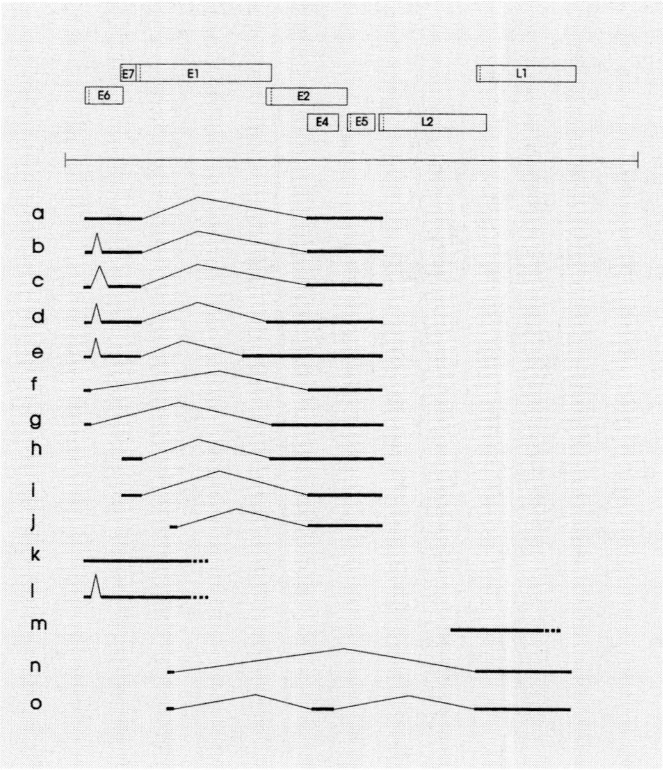

Fig. 1. Selected transcripts of HPV 16. The genome organization of HPV 16 is shown at the top. The slanted lines represent intervening sequences removed by splicing.

Transcripts

The activity of different promoters and complex splicing patterns result in an amazing variety of viral mRNAs. An example of this multitude is presented in figure 1, which shows structures of viral transcripts identified in different HPV-16-positive cells [5, 11, 15–17]. Many mRNAs contain several ORFs with translation initiation codons. There is evidence that translation of these polycistronic messages is influenced by alternative splicing. Little E7 protein appears to be translated from class a mRNAs (fig. 1), which primarily encode E6 since reinitiation of translation is unlikely because of the relative location of the end of E6 and the start of E7 [18]. In contrast, E7 can be expressed from spliced mRNA species b and c (fig. 1), because translation of E6 runs into an early stop codon after the splice, which increases the likelihood of reinitiation. This strategy to express oncoproteins E6 and E7 is characteristic for high-risk genital HPV types 16 and 18. The low-risk types 6 and 11 use a separate, E6-resident promoter to transcribe an E7-specific RNA [5].

The viral transcription patterns change considerably in different layers of the epithelium. There is only very weak expression of early genes in the basal layers of low-grade cervical dysplasias induced by HPV 16 or 33 and in some HPV-6- or HPV-11-induced condylomas [19–22]. Interestingly, however, the majority of HPV-6-containing condylomas tested by in situ hybridization displayed clear E6 signals within the basal layer as well as E7 signals in the 2–3 lowest layers of the epithelium (fig. 2A) [23]. One may speculate that expression of highly oncogenic E6/E7 proteins of HPV 16 became more restricted in evolution 'in order to' prevent neoplastic conversion of the cells which would be counterproductive from the 'standpoint of the virus'.

No E6 or E7 transcripts could be demonstrated in differentiated layers of many condylomas [23], and in HPV-16-induced lesions only single E7-expressing cells could be disclosed in the upper strata (fig. 2C) [24]. Several new early transcripts appear in the more differentiated cells where vegetative viral DNA replication takes place [19–24]. ORF E1- and E2-specific probes generate signals mostly restricted to nuclei which may suggest inefficient transport to the cytoplasm and limited translation. Abundant cytoplasmic mRNA species in the middle and upper parts of the epithelium of benign tumors probably correspond to the spliced E1/E4 transcript (class i, fig. 1). Transcripts coding for the late, structural proteins of the virus (classes m–o, fig. 1) could only be detected in the terminally differentiated keratinocytes of the superficial strata [20, 22, 24, 25]. As demonstrated by immunostaining, the histological localization of abundant HPV 16 transcripts correlates well with the presence of both E4 and L1 proteins [26].

In comparison to mild dysplasias, the E6/E7 region is actively transcribed in high-grade squamous intraepithelial lesions. In situ hybridization reveals evenly distributed signals throughout the undifferentiated epithelium [20, 21, 25]. There are essentially no differences in transcription patterns between these precancerous lesions and carcinomas (fig. 2B, D) which argues for transcriptional deregulation occurring rather early during tumor progression. In genital cancers, in situ hybridization failed to detect unspliced E6/E7 transcripts with the potential to express E6 protein [24], indicating that the overall low level of E6 mRNA [5] is due to low expression in every cell. Consequently, the observed increase in E6/E7 transcripts must result from elevated levels of mRNAs with a coding capacity for E7.

Abundant RNAs can be detected in many HPV-16-positive high-grade lesions and cancers with probes start-

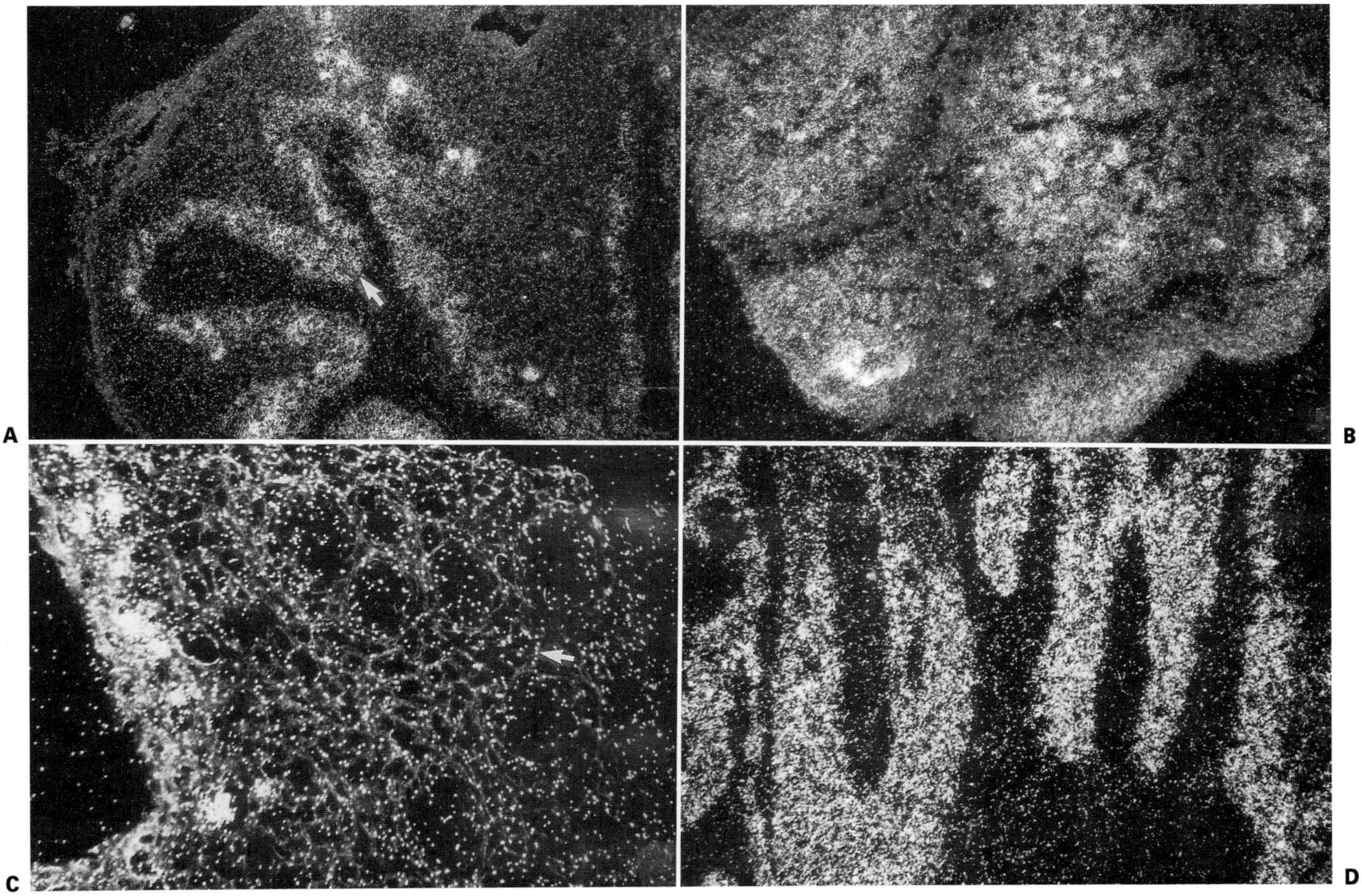

Fig. 2. In situ hybridization of tissue sections of a genital condyloma (**A**) and a cervical carcinoma (**B**), associated with HPV 6 and of a VIN I lesion (**C**) and a vulvar carcinoma (**D**), associated with HPV 16. The in situ hybridization was carried out with antisense riboprobes specific for HPV 6 E6 and E7 (HPV 6 nt 242–534; **A**, **B**) and HPV 16 E6/E7 (HPV 16 nt 183–551; **C**, **D**) messenger RNAs. White arrows indicate the position of the epithelial basement membrane. (Courtesy of T. Iftner, Erlangen, Germany).

ing within the E7 ORF and extending into E1 [24]. The signals obtained most probably correspond to the spliced E1/E4 message (class i, fig. 1) [11] since comparably strong signals can be seen after hybridization with a probe specific for the second exon. There were no second-exon signals in some cancers, which may reflect integration of viral DNA in these tumors, accompanied by disruption of the early transcription unit. In two cases of HPV-6-associated cancers, no E6 mRNA could be detected and the major transcript found corresponded to an E7 message [27].

In contrast to HPV-16-positive tumors, cancers associated with HPV 18 infections as a rule display interruption of the early transcription unit around the boundary of the E1 and E2 ORFs [16, 20, 21, 24, 25]. This reflects the different tendencies of both viruses to integrate into the host genome. Although the early region of HPV 16 appears to be intact in many cases, there is little or no evidence of mRNAs encoding a full-length E2 protein [16, 21, 24], which may reflect a specific down-regulation. E2-specific signals could be observed in HPV-6-associated cervical and urinary bladder cancers [27].

The late L1 and L2 transcripts generally disappear in the course of tumor progression but L1 RNAs are occasionally seen even in carcinomas [20, 21, 24]. It seems possible that the latter transcripts colinear to L1 arise from cellular promoters in sequences flanking integrated viral DNA [21]. Antisense transcripts detected in some cancers with a high viral DNA copy number are most probably of similar origin [28, 29].

Transcriptional Regulation by Viral E2 Protein

The papillomavirus ORF E2 encodes multifunctional phosphoproteins acting both as viral transcription modulators and replication factors [30, 31]. The E2 polypeptides of different virus types display a common structure consisting of conserved N- and C-terminal parts with roughly 220 and 95 amino acid residues, respectively, joined by a nonconserved central region of variable length. This setup correlates with different functional domains of the E2 proteins. The N-terminal moiety could be shown to be essential for trans-activating properties whereas the C-terminus mediates DNA binding and the ability to form dimer molecules. The central part of the E2 molecule is believed to form a flexible structure and is thus frequently referred to as the 'hinge' region. Comparison of primary amino acid sequences and structural features of the functional domains in the E2 polypeptides with those of other known transcription factors revealed only very little or no similarities, which suggests that E2 constitutes a prototype of a new class of regulatory proteins.

The E2 proteins interact with DNA as preexisting dimers [32–34]. The contact sites of each monomer reside within the major groove and are positioned on the same side of the DNA helix [32]. Interestingly, the binding event may result in bending of the DNA [35]. The E2-binding sites (E2-BSs) are palindromic sequences $ACC(N)_6GGT$ or $ACC(N)_6GTT$ [32, 33, 36, 37]. Multiple copies of these motifs occur in the genomes of all papillomaviruses, mostly clustered within the NCRs. The affinity of E2 proteins for these sites may vary by several orders of magnitude depending on differences in the nonconserved core of the recognition sites and in flanking sequences. The localization of the binding palindromes within the NCRs is partially conserved among papillomaviruses of the same phylogenetic lineages [38, 39].

Activation of transcription can be mediated only by the full-length forms of the E2 proteins [30, 31]. Whereas a single E2-BS mediates only moderate transactivation of a test promoter in the presence of E2, two closely located E2-BSs form a strong E2-dependent enhancer [40–42]. The synergistic effect of two adjacent E2-BSs relies in part on the cooperative binding of two E2 dimers [43]. However, there may be some additional mechanisms involved, e.g. the postulated cooperative interactions of two E2 dimers with other transcription factors such as USF or SP1 [44]. SP1 seems to be of special interest in this regard since it may form a physical complex with E2 and can stimulate E2 interactions with binding sites of weak affinity [45]. The molecular target(s) of E2 activation are not yet unequivocally identified, but evidence is accumulating in favor of components of the transcription preinitiation complex [46, 47].

Transactivation can occur over a large distance between the E2-BS and the target promoter, e.g. the P_{2443} promoter of BPV 1 may be activated via the E2 palindromes located well over 2 kb away within the viral NCR [48]. A single E2-BS proximal to a test promoter was shown to increase its E2 transactivation by other E2 sites located 1 kb away [41]. This can be nicely explained by the ability of distantly bound E2 molecules to interact directly and consequently loop the interspacing DNA sequences, as visualized by electron microscopy [34].

The E2 proteins not only activate, but also repress transcription by several mechanisms. In the BPV 1 system, N-terminally truncated forms of E2 have been demonstrated, which retain only the DNA-binding and dimerization domain [49–51]. They are expressed from mRNAs spliced to an acceptor site within E2 or initiated at the intragenic E2 promoter P_{3080}. These E2 repressor molecules may either 'block' the binding of the full-length E2 activator or form biologically inactive heterodimers with the latter molecules [52]. In BPV-1 transformed cells, the repressor forms of E2 actually dominate over the full-length protein [53] which is assumed to be a reason for the low levels of viral transcription observed. There are also indications for similar N-terminally truncated E2 forms in cells infected by HPV 11 and 16 [11, 17, 54, 55].

Another mechanism accounts for down-regulation of the P_{7185} promoter of BPV 1 by both full-length and N-terminal-truncated forms of E2 [56, 57]. In this case, E2 binding to its cognate palindrome downstream of the promoter obviously interferes with DNA binding or activity of a cellular factor which is essential for P_{7185} activity.

In HPV 11, 16 and 18, the full-length E2 acts as a repressor of the early promoters in front of ORF E6 [58–60]. The negative regulatory effect of E2 seems to be a consequence of the unique topography of protein-binding sites in the immediate vicinity of the basal promoter, observed in all genital HPV types. A tandem of two E2-BSs is flanked by an SP1-binding site and the TATA box element at a distance of 1 and 2–4 bp, respectively [61]. Down-regulation of the early promoter most probably occurs due to mutually exclusive binding of E2 and SP1 [61, 62] and/or E2 and TATA-box-binding protein [46, 59, 60], which should impede the SP1-mediated activation and/or the formation of the transcription preinitiation complex at the TATA box.

Relatively little is known about the regulatory effects of E2 in the group of skin-specific HPV types. In the epidermodysplasia-verruciformis-associated HPV 8, E2 activates the late viral promoter P_{7535} [6]. Recent results showed that this effect is mainly due to E2 binding within the transcriptional silencer located upstream of the promoter, which overrides the negative regulatory function [6, 63]. These observations suggest a possible role of E2 in the switch between early and late gene transcription.

Transcription Control by Cellular Regulatory Proteins

Most cis-regulatory elements of papillomaviruses have been mapped to the NCR. Depending on the papillomavirus type, the NCRs roughly extend over 400–900 bp [64, 65]. In all papillomaviruses studied in this regard, the NCRs are heavily involved in sequence-specific interactions with cellular proteins [66–68]. Adding to the structural and functional complexity of the NCR sequences, the origins of viral DNA replication (ori) could also be shown to reside within these regions [69–71].

Functional dissection of reguatory sequences of HPV 11 [58, 72], 16 [73–75] and 18 [76, 77] revealed transcriptional enhancers in the central part of the NCRs. They are composed of a number of sequence modules of low or moderate activity which cooperate to generate a strong enhancer [39, 78, 79]. Its activity depends on interactions with cellular transcription factors such as jun/fos [78, 80, 81], NF1 [78, 82, 83], octamer-binding proteins [75, 84, 85], TEF 1 and TEF 2 [75, 79]. A similar set of binding sites for cellular proteins is generally found in the center of the NCRs of genital HPVs [39]. Both AP1 and NF1 recognition sites are consistent features of all enhancers. The exact constellation of the various protein recognition sequences differs, however, which is in line with the concept of a modular setup.

An intriguing functional aspect of the main enhancer in genital HPVs is its keratinocyte specificity [58, 73, 74, 76] which may represent at least one basis for the tissue tropism of these viruses. Several studies suspected factors junB, AP1, NF1 or TEF 1/TEF 1 cofactor of contributing to keratinocyte dependence [79–81, 83]. AP1 and NF1 are widely regarded as ubiquitous transcription factors but actually represent heterogeneous protein families, and relatively little is known about the cell specificity of individual members. It is quite possible that the cell specificity arises from subtle differences in the balance of several transcription factors. There is so far only one example of a genuine epithelial enhancer factor (KRF-1) which binds to the enhancer of HPV 18 [86] but KRF-1-binding sites could not be identified in other genital HPVs. In view of still undefined cellular proteins interacting with the main enhancer, the question for the molecular basis of the tissue specificity must still be regarded as open.

All genital HPV types contain within their NCRs elements which mediate positive regulation of viral transcription by glucocorticoid hormones like progesterone [74, 87]. In HPV 16, these sequences could be identified both within the central enhancer as well as in its flanking regions. Their activity may be presumed to be especially important for the biology of the genital HPV infections: expression of viral functions may fluctuate with the temporal hormonal status of the host (e.g. in the course of an ovulation cycle or pregnancy), and the use of hormonal contraceptives may result in increased levels of the viral oncoproteins in the HPV-infected cervix. It is noteworthy in this context that progesterone or progestin from oral contraceptives enhanced the transformation of rodent cells by HPV 16 [88], and dexamethasone increased the transformation rate of HPV-16-immortalized keratinocytes by the Ha-*ras* oncogene [89].

A second type of constitutive enhancer was disclosed within the L1 proximal part of the NCR in HPV 8, 11, 18 and BPV 1 [90–94], whose activity is independent of the host cell type in contrast to the main enhancers of genital HPVs. The HPV 8 enhancer activates the late viral promoter [6, 90].

The papillomavirus NCRs were more recently shown to contain elements which down-regulate the activity of viral promoters. These findings are in line with some earlier observations suggesting that the viral regulatory sequences are actually composed of a mosaic of positive and negative control motifs [95, 96]. Silencers of transcription have so far been identified for HPVs 8, 16 and 18. One type of negative control element of genital HPVs 16 and 18 depends for its activity on the interaction of viral sequences with the multifunctional regulatory factor yin-yang 1 (YY1) [97, 98]; others rely on the retinoic acid receptor [99] or the NF-IL6 factor [100]. In contrast, the recently described silencer of the HPV 8 late promoter seems to be independent of interactions with cellular proteins [63].

The repression by YY1 seems to be biologically particularly relevant. Deletions of 2–4 YY1 binding sites were detectable in the HPV 16 NCRs from three out of six cervical cancers with purely episomal viral DNA. A seventh case showed a point mutation in one YY1 recognition site that prevented binding [98, 101]. All these mutations resulted in a 3.5- to 6.5-fold increased activity of the oncogene promoter P_{97}. Elimination of YY1 target sequences of HPV 16 thus seems to be a way taken repeatedly to escape cellular repression. The consequently enhanced expression of viral oncoproteins may be a crucial step in tumor progression. A deletion of 38 bp from the HPV 16 NCR in SiHa cells [102] removes one of the identified YY1-binding sites [98] which may indicate that inactivation of the YY1-dependent repression system is not restricted to episomal HPV DNA in cancers.

Cotransfection of expression vectors for the cellular transcription factor Oct-1 with HPV 18-NCR-reporter gene constructs demonstrated repression of the HPV 18 enhancer by Oct-1 [103]. The down-modulation turned out to be independent of Oct-1 binding to DNA, which suggests repression due to squelching. This would be another interesting mechanism of negative control although the effect may be less pronounced with physiological Oct-1 concentrations.

This complicated network of positive and negative control mechanisms is obviously sensitive to extracellular signals. The cytokines transforming growth factor (TGF)-β_1 and -β_2 [104, 105], epidermal growth factor (EGF) [106], leukoregulin, interferon-γ [107], interleukin-1 [108] and tumor necrosis factor (TNF)-α [108, 109] all suppress the expression of HPV E6 and E7 genes. Most of them repress viral transcription selectively in nonmalignant cells indicating that signal transmission is interrupted in cancer cells. It is not yet known which positively or negatively acting transcription factors are targets of intracellular signal transmission pathways to finally mediate repression of HPV transcription.

A series of recent observations identified a function encoded on chromosome 11 and protein phosphatase 2A (PP2A) as intracellular switch points in the regulation of HPV transcription. A differential control was described in nonmalignant hybrids of cervical cancer cells with normal keratinocytes and tumorigenic segregants depending on the presence of a normal chromosome 11 [110, 111]. Viral transcription was suppressed in nontumorigenic hybrids after heterografting into nude mice but was retained in segregants which lacked one normal chromosome 11 [112]. HPV 16 is virtually inactive in diploid human fibroblasts but efficiently transcribed in fibroblasts with a deletion in the short arm of chromosome 11, which are distinguished by high levels of the 55-kDa repressor subunit of PP2A and by a consequently low activity of PP2A [113]. In normal fibroblasts, inhibition of the phosphatase by okadaic acid resulted in a similar stimulation of HPV transcription. The inhibition of PP2A certainly exerts mostly pleiotropic effects. Among others, it appears to affect proteins interacting with the TATAAA motif of the HPV 16 promoter [114]. These data identified three steps of negative control in a chain of events which starts with the still uncharacterized function encoded on chromosome 11, directly or indirectly inhibiting the phosphatase inhibitor, and proceeds with the certainly indirect inhibition of the initiation of viral transcription by PP2A. Both the chromosome 11 function and PP2A fulfil criteria of a cellular interfering factor(s) postulated by zur Hausen [115, 116] to explain intracellular surveillance of oncogenic papillomaviruses by the host.

References

1 Baker CC: Bovine papillomavirus type 1 transcription; in Pfister H (ed): Papillomaviruses and Human Cancer. Boca Raton, CRC, 1990; 91–112.

2 Furth PA, Baker CC: An element in the bovine papillomavirus late 3′ untranslated region reduces polyadenylated cytoplasmic RNA levels. J Virol 1991;65:5806–5812.

3 Smotkin D, Wettstein FO: Transcription of human papillomavirus type 16 early genes in a cervical cancer and a cancer derived cell line and identification of the E7 protein. Proc Natl Acad Sci USA 1986;83:4680–4684.

4 Thierry F, Heard JM, Dartmann K, Yaniv M: Characterization of a transcriptional promoter of human papillomavirus 18 and modulation of its expression by simian virus 40 and adenovirus early antigens. J Virol 1987;61:134–142.

5 Smotkin D, Prokoph H, Wettstein FO: Oncogenic and nononcogenic human genital papillomaviruses generate the E7 mRNA by different mechanisms. J Virol 1989;63:1441–1447.

6 Stubenrauch F, Malejczyk J, Fuchs PG, Pfister H: Late promoter of human papillomavirus 8 and its regulation. J Virol 1992;66:3485–3493.

7 Iftner T, Sagner G, Pfister H, Wettstein FO: The E7 protein of human papillomavirus 8 is a nonphosphorylated protein of 17 kDa and can be generated by two different mechanisms. Virology 1990;179:428–436.

8 Karlen S, Beard P: Identification of novel promoters in the genome of human papillomavirus type 18. J Virol 1993;67:4296–4306.

9 Chow LT, Reilly SS, Broker TR, Taichman LB: Identification and mapping of human papillomavirus type 1 RNA transcripts recovered from plantar warts and infected epithelial cell cultures. J Virol 1987;61:1913–1918.

10 Nasseri M, Hirochika R, Broker TR, Chow LT: A human papillomavirus type 11 transcript encoding an E1-E4 protein. Virology 1987;159: 433–439.

11 Doorbar J, Parton A, Hartley L, Crook T, Stanley M, Crawford L: Detection of novel splicing patterns in a HPV16-containing keratinocyte cell line. Virology 1990;178:254–262.

12 Palermo-Dilts DA, Broker TR, Chow LT: Human papillomavirus 1 produces redundant as well as polycistronic mRNAs in plantar warts. J Virol 1990;64:3144–3149.

13 Wettstein FO: State of viral DNA and gene expression in benign vs. malignant tumors; in Pfister H (ed): Papillomaviruses and Human Cancer. Boca Raton, CRC, 1990, pp 155–179.
14 Chow LT, Nasseri M, Wolinski SM, Broker TR: Human papillomavirus types 6 and 11 mRNAs from genital condylomata acuminata. J Virol 1987;61:2581–2588.
15 Rohlfs M, Winkenbach S, Meyer S, Rupp T, Dürst M: Viral transcription in human keratinocyte cell lines immortalized by human papillomavirus type-16. Virology 1991;183:331–342.
16 Sherman L, Alloul N, Golan I, Dürst M, Baram A: Expression and splicing pattern of human papillomavirus type-16 mRNAs in pre-cancerous lesions and carcinomas of the cervix, in human keratinocytes immortalized by HPV16 and in cell lines established from cervical cancers. Int J Cancer 1992;50:356–364.
17 Sherman L, Alloul N: Human papillomavirus type 16 expresses a variety of alternatively spliced mRNAs putatively encoding the E2 protein. Virology 1992;191:953–959.
18 Sedman AS, Barbosa M, Vass WC, Hubbert NL, Haas JA, Lowy DR, Schiller JT: The full-length E6 protein of human papillomavirus 16 has transforming and trans-activating activities and cooperates with E7 to immortalize keratinocytes in culture. J Virol 1991;65:4860–4866.
19 Beyer-Finkler E, Stoler MH, Girardi F, Pfister H: Cell differentiation-related gene expression of human papillomavirus 33. Med Microbiol Immunol 1990;179:185–192.
20 Dürst M, Glitz D, Schneider A, zur Hausen H: Human papillomavirus type 16 (HPV16) gene expression and DNA replication in cervical neoplasia: Analysis by in situ hybridization. Virology 1992;189:132–140.
21 Stoler MH, Rhodes CR, Whitbeck A, Wolinsky SM, Chow LT, Broker TR: Human papillomavirus type 16 and 18 gene expression in cervical neoplasias. Hum Pathol 1992;23:117–128.
22 Stoler MH, Wolinsky SM, Whitbeck A, Broker TR, Chow LT: Differentiation-linked human papillomavirus types 6 and 11 transcription in genital condylomata revealed by in situ hybridization with message-specific RNA probes. Virology 1989;172:331–340.
23 Iftner T, Oft M, Böhm S, Wilczynski SP, Pfister H: Transcription of the E6 and E7 genes of human papillomavirus type 6 in anogenital condylomata is restricted to undifferentiated cell layers of the epithelium. J Virol 1992;66:4639–4646.
24 Böhm S, Wilczynski SP, Pfister H, Iftner T: The predominant mRNA class in HPV16-infected genital neoplasias does not encode the E6 or the E7 protein. Int J Cancer 1993;55:791–798.
25 Higgins GD, Uzelin DM, Phillips GE, McEvoy P, Marin R, Burrell CJ: Transcription patterns of human papillomavirus type 16 in genital intraepithelial neoplasia: Evidence for promoter usage within E7 open reading frame during epithelial differentiation. J Gen Virol 1992;73:2047–2057.
26 Crum CP, Barber S, Symbula M, Snyder K, Saleh AM, Roche JK: Coexpression of the human papillomavirus type 16 E4 and L1 open reading frames in early cervical neoplasia. Virology 1990;178:238–246.
27 Oft M, Böhm S, Wilczynski SP, Iftner T: Expression of the different viral mRNAs of human papilloma virus 16 in a squamous-cell carcinoma of the bladder and the cervix. Int J Cancer 1993;53:924–931.
28 Higgins GD, Uzelin DM, Phillips GE, Burrell CJ: Presence and distribution of human papillomavirus sense and antisense RNA transcripts in genital cancers. J Gen Virol 1991;72:885–895.
29 Vormwald-Dogan V, Fischer B, Bludau H, Freese UK, Gissmann L, Glitz D, Schwarz E, Dürst M: Sense and antisense transcripts of human papillomavirus type 16 in cervical cancers. J Gen Virol 1992;73:1833–1838.
30 McBride AA, Romanczuk H, Howley PM: The papillomavirus E2 regulatory proteins. J Biol Chem 1991;266:18411–18414.
31 Steger G, Ham J, Thierry F, Yaniv M: The E2 proteins of papillomaviruses as transcription and replication factors; in Doerfler W, Böhm P (eds): Virus Strategies: Molecular Biology and Pathogenesis. Weinheim, Verlag Chemie, 1993, pp 263–278.
32 Dostatni N, Thierry F, Yaniv M: A dimer of BPV-1 E2 containing a protease resistant core interacts with its DNA targets. EMBO J 1988; 7:3807–3816.
33 McBride AA, Byrne JC, Howley PM: E2 polypeptides encoded by the bovine papillomavirus type 1 form dimers through the common carboxyl-terminal domain: Transactivation is mediated by the conserved amino-terminal domain. Proc Natl Acad Sci USA 1989;86:510–514.
34 Knight J, Li R, Botchan M: The activation domain of the bovine papillomavirus E2 protein mediates association of DNA-bound dimers to form DNA loops. Proc Natl Acad Sci USA 1991;88:3204–3208.
35 Moskaluk C, Bastia D: DNA bending is induced in an enhancer by the DNA-binding domain of the bovine papillomavirus E2 protein. Proc Natl Acad Sci USA 1988;85:1826–1830.
36 Androphy E, Lowy DR, Schiller JT: Bovine papillomavirus E2 transactivating gene product binds to specific sites in papillomavirus DNA. Nature 1987;325:70–73.
37 Li R, Knight J, Bream G, Stenlund A, Botchan M: Specific recognition nucleotides and their DNA context determine the affinity of E2 protein for 17 binding sites in the BPV1 genome. Genes Dev 1989;3:510–526.
38 Ensser A, Pfister H: Epidermodysplasia verruciformis associated human papillomaviruses present a subgenus-specific organization of the regulatory genome region. Nucleic Acids Res 1990;18:3919–3922.
39 Chong T, Chan WK, Bernard HU: Transcriptional activation of human papillomavirus 16 by nuclear factor I, AP1, steroid receptors and a possible novel transcription factor, PVF: A model for the composition of genital papillomavirus enhancers. Nucleic Acids Res 1990; 18:465–470.
40 Hawley-Nelson P, Androphy EJ, Lowy DR, Schiller JT: The specific DNA recognition sequence of the bovine papillomavirus E2 protein is an E2-dependent enhancer. EMBO J 1988;7:525–531.
41 Thierry F, Dostatni N, Arnos F, Yaniv M: Cooperative activation of transcription by bovine papillomavirus type 1 E2 can occur over a large distance. Mol Cell Biol 1990;10:4431–4437.
42 Gauthier JM, Dostatni N, Lusky M, Yaniv M: Two DNA-bound E2 dimers are required for strong transcriptional activation and for cooperation with cellular factors in most cells. New Biol 1991;3:1–12.
43 Monini P, Grossman SR, Peppinsky B, Androphy EJ, Laimins LA: Cooperative binding of the E2 protein of bovine papillomavirus to adjacent E2-responsive sequences. J Virol 1991;65:2124–2130.
44 Ham J, Dostatni N, Arnos F, Yaniv M: Several different upstream promoter elements can potentiate transactivation by the BPV-1 E2 protein. EMBO J 1991;10:2931–2940.
45 Li R, Knight JD, Jackson SP, Tijan R, Botchan M: Interaction between Sp1 and the BPV enhancer E2 protein mediates synergistic activation of transcription. Cell 1991;65:493–505.
46 Dostatni N, Lambert PF, Sousa R, Ham J, Howley PM, Yaniv M: The E2 protein can act as a repressor by preventing formation of the initiation complex. Genes Dev 1991;5:1657–1671.
47 Ham J, Steger G, Yaniv M: Cooperativitiy in vivo between the E2 transactivator and the TATA box binding protein depends on core promoter structure. EMBO J 1994;13:147–157.
48 Spalholz BA, Vande Pol SB, Howley PM: Characterization of the cis-elements involved in basal and E2-transactivated expression of the bovine papillomavirus P_{2443} promoter. J Virol 1991;65:743–753.
49 Lambert PF, Spalholz BA, Howley PM: A transcriptional repressor encoded by BPV-1 shares a common carboxy-terminal domain with the E2 transactivator. Cell 1987;50:69–78.
50 Lambert PF, Hubbert NL, Howley PM, Schiller JT: Genetic assignment of multiple E2 gene products in bovine papillomavirus-transformed cells. J Virol 1989;63:3151–3154.
51 Choe J, Vaillancourt P, Stenlund A, Botchan M: Bovine papillomavirus type 1 encodes two forms of a transcriptional repressor: Structural and functional analysis of new viral cDNAs. J Virol 1989;63:1743–1755.
52 Barsoum J, Prakash SS, Han P, Androphy EJ: Mechanism of action of the papillomavirus E2 repressor: Repression in the absence of DNA binding. J Virol 1992;66:3941–3945.

53 Lambert PF, Monk BC, Howley PM: Phenotypic analysis of bovine papillomavirus type 1 E2 repressor mutants. J Virol 1990;64:950–956.
54 Chin MT, Hirochika R, Hirochika H, Broker TR, Chow LT: Regulation of human papillomavirus type 11 enhancer and E6 promoter by activating and repressing proteins from the E2 open reading frame: Functional and biochemical studies. J Virol 1988;62:2994–3002.
55 Chiang CM, Broker TR, Chow LT: An E1ME2C fusion protein encoded by human papillomavirus type 11 is a sequence-specific transcription repressor. J Virol 1991;65:3317–3329.
56 Vande Pol SB, Howley PM: A bovine papillomavirus constitutive enhancer is negatively regulated by the E2 repressor through competitive binding for a cellular factor. J Virol 1990; 64:5420–5429.
57 Stenlund A, Botchan M: The E2 transactivator can act as a repressor by interfering with a cellular transcription factor. Genes Dev 1990;4: 123–136.
58 Dollard SC, Broker TR, Chow LT: Regulation of the human papillomavirus type 11 E6 promoter by viral and host transcription factors in primary human keratinocytes. J Virol 1993;67: 1721–1726.
59 Thierry F, Howley PM: Functional analysis of E2-mediated repression of the HPV18 P_{105} promoter. New Biologist 1991;3:90–100.
60 Romanczuk N, Thierry F, Howley PM: Mutational analysis of cis elements involved in E2 modulation of human papillomavirus P_{97} and type 18 P_{105} promoters. J Virol 1990;64:2849–2859.
61 Tan SH, Gloss B, Bernard HU: During negative regulation of the human papillomavirus-16 E6 promoter, the viral E2 protein can displace Sp1 from a proximal promoter element. Nucleic Acids Res 1992;20:251–256.
62 Gloss B, Bernard HU: The E6/E7 promoter of human papillomavirus type 16 is activated in the absence of E2 proteinby sequence-aberrant Sp1 distal element. J Virol 1990;64:5577–5584.
63 May M, Grassmann K, Pfister H, Fuchs PG: Transcriptional silencer of the human papillomavirus type 8 late promoter interacts alternatively with the viral transactivator E2 or with a cellular factor. J Virol 1994;68:3612–3619.
64 Pfister H, Fuchs PG: Papillomaviruses: Particles, genome organization and proteins; in Syrjänen K, Gissmann L, Koss L (eds): Papillomaviruses and Human Disease. Heidelberg, Springer, 1987, pp 1–22.
65 Iftner T: Papillomavirus genomes: Sequence analysis related to functional aspects; in Pfister H (ed): Papillomaviruses and Human Cancer, Boca Raton, CRC, 1990, pp 181–202.
66 Garcia-Carranca A, Thierry F, Yaniv M: Interplay of viral and cellular proteins along the long control region of human papillomavirus type 18. J Virol 1988;62:4321–4330.
67 Gloss B, Chong T, Bernard HU: Numerous nuclear proteins bind the long control region of human papillomavirus type 16: A subset of 6 of 23 DNase I-protected segments coincides with the location of the cell-type-specific enhancer. J Virol 1989;63:1142–1152.
68 May M, Helbl V, Pfister H, Fuchs PG: Unique topography of DNA-protein interactions in the non-coding region of epidermodysplasia verruciformis-associated human papillomaviruses. J Gen Virol 1991;72:2989–2997.
69 Ustav M, Stenlund A: Identification of the origin of replication of bovine papillomavirus and characterization of the viral origin recognition factor E1. EMBO J 1991;10:4321–4329.
70 Chiang CM, Ustav M, Stenlund A, Ho TF, Broker TR, Chow LT: Viral E1 and E2 proteins support replication of homologous and heterologous papillomavirus origins. Proc Natl Acad Sci USA 1992;89:5799–5803.
71 Del Vecchio AM, Romanczuk H, Howley PM, Baker CC: Transient replication of human papillomavirus DNAs. J Virol 1992;66:5949–5958.
72 Chin MT, Broker TR, Chow LT: Identification of a novel constitutive enhancer element and an associated binding protein: Implications for human papillomavirus type 11 regulation. J Virol 1989;63:2967–2976.
73 Cripe TP, Haugen TH, Turk JP, Tabatabai F, Schmid PG III, Dürst M, Gissmann L, Roman A, Turek L: Transcriptional regulation of the human papillomavirus-16 promoter by a keratonocyte-dependent enhancer, and by viral E2 trans-activator and repressor gene products: Implications for cervical carcinogenesis. EMBO J 1987;6:3745–3753.
74 Gloss B, Bernard HU, Seedorf K, Klock K: The upstream regulatory region of the human papillomavirus-16 contains an E2 protein-independent enhancer which is specific for cervical carcinoma cells and regulated by glucocorticoid hormones. EMBO J 1987;6:3735–3743.
75 Chong T, Apt D, Gloss B, Isa M, Bernard HU: The enhancer of human papillomavirus type 16: Binding sites for the ubiquitous transcription factors oct-1, NFA, TEF-2, NF1, and AP-1 participate in epithelial cell-specific transcription. J Virol 1991;65:5933–5943.
76 Swift FV, Bhat K, Younghusband HB, Hamada H: Characterization of a cell type-specific enhancer found in the human papilloma virus type 18 genome. EMBO J 1987;6:1339–1344.
77 Gius D, Grossman S, Bedell MA, Laimins LA: Inducible and constitutive enhancer domains in the noncoding region of human papillomavirus type 18. J Virol 1988;62:665–672.
78 Cripe TP, Alderborn A, Anderson RD, Parkkinen S, Bergman P, Haugen TH, Petterson U, Turek LP: Transcriptional activation of the human papillomavirus-16 P97 promoter by an 88-nucleotide enhancer containing distinct cell-dependent and AP1-responsive modules. New Biologist 1990;2:450–463.
79 Ishiji T, Lace MJ, Pakkinen S, Anderson RD, Haugen TH, Cripe TP, Xiao JH, Davidson I, Chambon P, Turek LP: Transcriptional enhancer factor (TEF)-1 and its cell-specific cofactor activate human papillomavirus-16 E6 and E7 oncogene transcription in keratinocytes and cervical carcinoma cells. EMBO J 1992;11: 2271–2281.
80 Offord EA, Beard P: A member of the activator protein 1 family found in keratinocytes but not in fibroblasts required for transcription from a human papillomavirus type 18 promoter. J Virol 1990;64:4792–4798.
81 Thierry F, Spirou G, Yaniv M, Howley PM: Two AP1 sites binding *jun* B are essential for human papillomavirus type 18 transcription in keratinocytes. J Virol 1992;66:3740–3748.
82 Gloss B, Yeo-Gloss M, Meisterernst M, Rogge L, Winnacker EL, Bernard HU: Clusters of nuclear factor 1 binding sites identify enhancers of several papillomaviruses but alone are not sufficient for enhancer function. Nucleic Acids Res 1989;17:3519–3533.
83 Apt D, Chong T, Liu Y, Bernard HU: Nuclear factor 1 and epithelial cell-specific transcription of human papillomavirus type 16. J Virol 1993;67:4455–4463.
84 Royer HD, Freyaldenhoven MP, Napierski I, Spitkovsky DD, Bauknecht T, Dathan N: Delineation of human papillomavirus type 18 enhancer binding proteins: The intracellular distribution of a novel octamer binding protein p92 is cell cycle regulated. Nucleic Acids Res 1991;19:2363–2371.
85 Morris PJ, Dent CL, Ring CJ, Latchman DS: The octamer binding site in the HPV16 regulatory region produces opposite effects on gene expression in cervical and non-cervical cells. Nucleic Acids Res 1993;21:1019–1023.
86 Mack DH, Laimins LA: A keratinocyte-specific transcription factor, KRF-1, interacts with AP-1 to activate of human papillomavirus type 18 in squamous epithelial cells. Proc Natl Acad Sci USA 1991;88:9102–9106.
87 Mittal R, Pater A, Pater MM: Multiple human papillomavirus type 16 glucocorticoid response elements functional for transformation, transient expression, and DNA-protein interactions. J Virol 1993;67:5656–5659.
88 Pater A, Bayatpour M, Pater MM: Oncogenic transformation by a human papillomavirus 16 deoxyribonucleic acid in the presence of progesterone or progestins from oral contraceptives. Am J Obstet Gynecol 1990;162:1099–1103.
89 Dürst M, Gallahan D, Jay G, Rhin JS: Glucocorticoid-enhanced neoplastic transformation of human keratinocytes by human papillomavirus type 16 and an activated *ras*-oncogene. Virology 1989;173:767–771.
90 Horn S, Pfister H, Fuchs PG: Constitutive transcriptional activator of epidermodysplasia verruciformis-associated human papillomavirus 8. Virology 1993;196:674–681.
91 Auborn KJ, Galli RL, Dilorenzo TP, Steinberg BM: Identification of DNA-protein interactions and enhancer activity at the 5′ end of the upstream regulatory region in human papillomavirus type 11. Virology 1989;170:123–130.

92 Auborn KJ, Steinberg BM: A key DNA-protein interaction determines the function of the 5′ URR enhancer in human papillomavirus type 11. Virology 1991;181:132–138.
93 Hoppe-Seyler F, Butz K: A novel cis-stimulatory element maps to the 5′ portion of the human papillomavirus type 18 upstream regulatory region and is functionally dependent on a sequence-aberrant Sp1 binding site. J Gen Virol 1993;74:281–286.
94 Vande Pol SB, Howley PM: A bovine papillomavirus constitutive enhancer is negatively regulated by the E2 repressor through competitive binding for a cellular factor. J Virol 1990;64:5420–5429.
95 Reh H, Pfister H: Human papillomavirus type 8 contains cis-active positive and negative transcriptional control sequences. J Gen Virol 1990;71:2457–2462.
96 Jackson ME, Campo MS: Positive and negative E2-independent regulatory elements in the long control region of bovine papillomavirus type 4. J Gen Virol 1991;72:877–883.
97 Bauknecht T, Angel P, Royer HD, zur Hausen H: Identification of a negative regulatory domain in the human papillomavirus type 18 promoter: Interaction with the transcriptional repressor YY1. EMBO J 1992;11: 4607–4617.
98 May M, Dong XP, Beyer-Finkler E, Stubenrauch F, Fuchs PG, Pfister H: The E6-E7 promoter of extrachromosomal HPV16 DNA in cervical cancers escapes from cellular repression by mutation of target sequences for YY1. EMBO J 1994;13:1460–1466.
99 Bartsch D, Boye B, Baust C, zur Hausen H, Schwarz E: Retinoic acid-mediated repression of human papillomavirus 18 transcription and different ligand regulation of the retinoic acid receptor in non-tumorigenic and tumorigenic HeLa hybrid cells. EMBO J 1992;11:2283–2291.
100 Kyo S, Inoue M, Nishio Y, Nakanishi K, Akira S, Inoue H, Yutsudo M, Tanizawa O, Hakura A: NF-IL 6 represses early gene expression of human papillomavirus type 16 through binding to the noncoding region. J Virol 1993;67:1058–1066.
101 Dong XP, Stubenrauch F, Beyer-Finkler E, Pfister H: Prevalence of deletions of YY1 binding sites in episomal HPV 16 DNA from cervical cancers. Int J Cancer, in press.
102 Ho L, Chan SY, Chow V, Chong T, Tay SK, Villa LL, Bernard HU: Sequence variants of human papillomavirus type 16 in clinical samples permit verification and extension of epidemiological studies and construction of a phylogenetic tree. J Clin Microbiol 1991;29: 1765–1772.
103 Hoppe-Seyler F, Butz K, zur Hausen H: Repression of the human papillomavirus type 18 enhancer by the cellular transcription factor Oct-1. J Virol 1991;65:5613–5618.
104 Braun L, Dürst M, Mikumo R, Guipposo P: Differential response of nontumorigenic and tumorigenic human papillomavirus type 16-positive epithelial cells to transforming growth factor. Cancer Res 1990;50:7324–7332.
105 Woodworth CD, Notario V, DiPaolo JA: Transforming growth factor beta 1 and 2 transcriptionally regulate human papillomavirus (HPV) type 16 early gene expression in HPV-immortalized human genital epithelial cells. J Virol 1990;64:4767–4775.
106 Yasumoto S, Taniguchi A, Sohma K: Epidermal growth factor (EGF) elicits downregulation of human papillomavirus-type 16 (HPV16) E6/E7 mRNA at the transcriptional level in an EGF-stimulated human keratinocyte cell line: Functional role of EGF-responsive silencer in the HPV-16 long control region. J Virol 1991;65:2000–2009.
107 Woodworth CD, Lichti U, Simpson S, Evans CH, DiPaolo JA: Leukoregulin and gamma-interferon inhibit human papillomavirus type 16 gene transcription in human papillomavirus-immortalized human cervical cells. Cancer Res 1992;52:456–463.
108 Kyo S, Inoue M, Hayasaka N, Inoue T, Yutsudo M, Tanizawa O, Hakura A: Regulation of early gene expression of human papillomavirus type 16 by inflammatory cytokines. Virology 1994;200:130–139.
109 Rösl F, Lengert M, Albrecht J, Kleine K, Zawatzky R, Schraven B, zur Hausen H: Differential regulation of the JE gene encoding the monocyte chemoattractant protein (NCP-1) in cervical carcinoma cells and derived hybrids. J Virol 1994;68:2142–2150.
110 Rösl F, Dürst M, zur Hausen H: Selective suppression of human papillomavirus transcription in non-tumorigenic cells by 5-azacytidine. EMBO J 1988;7:1321–1328.
111 Rösl F, Achtstetter T, Hutter KJ, Bauknecht T, Futterman G, zur Hausen H: Extinction of the HPV18 upstream regulatory region in cervical carcinoma cells after fusion with non-tumorigenic human keratinocytes under non-selective conditions. EMBO J 1991;10:1337–1345.
112 Bosch F, Schwarz E, Boucamp P, Fusenig NE, Bartsch D, zur Hausen H: Suppression in vivo of human papillomavirus type 18 E6-E7 gene expression in nontumorigenic HeLa-fibroblast hybrid cells. J Virol 1990;64:4743–4754.
113 Smits PHM, Smits HL, Minnaar RP, Hemmings BA, Mayer-Jaekel RE, Schuurman T, van der Noordaa J, ter Schegget J: The 55kDa regulatory subunit of protein phosphatase 2A plays a role in the activation of the HPV16 long control region in human cells with a deletion in the short arm of chromosome 11. EMBO J 1992;11:4601–4606.
114 Smits PHM, Smits HL, Minnaar RP, ter Schegget J: Regulation of human papillomavirus type 16 (HPV-16) transcription by loci on the short arm of chromosome 11 is mediated by the TATAAAA motif of the HPV-16 promoter. J Gen Virol 1993;74:121–124.
115 zur Hausen: Cell-virus gene balance hypothesis of carcinogenesis. Behring Mitt 1977;61: 23–30.
116 zur Hausen H: Intracellular surveillance of persisting viral infections. Human genital cancer results from deficient cellular control of papillomavirus gene expression. Lancet 1986;ii:489–491.

Intervirology 1994;37:168–179

Hubert Stöppler
Melissa Conrad Stöppler
Richard Schlegel

Department of Pathology, Georgetown University Medical Center, Washington, D.C. (USA)

Transforming Proteins of the Papillomaviruses

Key Words
Papillomaviruses
Carcinogenesis
Viral transforming proteins
Tumor suppressor proteins (p53, Rb)
Transcription
Human epithelial cells
Primary rodent cells
Immortalized rodent cells
Growth factor receptors

Introduction

The papillomaviruses are widespread among vertebrate species and play a significant role in the induction of both benign and malignant neoplasia [1–9]. In humans, papillomaviruses have a critical role in the development of cervical cancer, a disease with enormous costs to the medical health care system. Understanding the role of the papillomavirus genes in neoplasia will not only provide insights into cell growth control but will also assist in the design of future approaches to preventing and inhibiting the occurrence and progression of this disease.

Many papillomavirus genomes encode three independent transforming proteins: E7, E6, and E5. The activity of these genes and corresponding proteins has been studied in several animal species, including human. The first papillomavirus to be cloned, sequenced and characterized for its transforming genes was the bovine papillomavirus type 1 (BPV1) [10–12]. Several years later, human papillomavirus (HPV) genomes were detected in and cloned from human genital tumors [5, 8] and their ability to immortalize and alter the differentiation of keratinocytes was discovered [13–15]. The development of in vitro assays for cellular transformation eventually led to the molecular characterization of HPV transforming genes in keratinocytes [15–23] and fibroblasts [24–30]. However, the first insight into the definition of potential HPV transforming genes was obtained from sequence analysis of HPV DNA in cervical cancers. It was appreciated early on, that cervical carcinomas contained HPV genomes which, although disrupted and rearranged as a consequence of integration, uniformly expressed the E6 and E7 genes [31, 32]. This was in contrast to several other early viral genes such as E1, E2, E4, and E5 which were often separated from the viral promoter, deleted, or significantly rearranged. The consistent retention and expression of the E6/E7 genes in tumors, as well as studies utilizing E6/E7 antisense plasmids [33], suggest that these genes are required not only for the development of the malignant state, but also for its maintenance.

In the following sections, we describe the properties of the papillomavirus transforming proteins with a primary focus on the E6/E7 genes of the human papillomaviruses. While we generically group the E7, E6, and E5 proteins as 'transforming proteins,' it must be appreciated that these proteins are incapable of converting primary human cells

Hubert Stöppler
Department of Pathology
Georgetown University Medical Center
3900 Reservoir Road
Washington, DC 20007 (USA)

0300–5526/94/0374–0168
$8.00/0

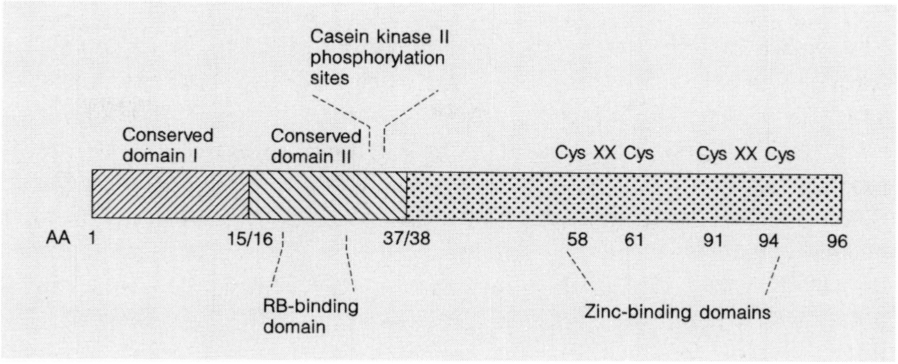

Fig. 1. Diagrammatic representation of the HPV 16 E7 protein. Numbers indicate amino acid (AA) positions. Conserved domains (CD) I and II are regions with homologies to the adenovirus E1A and the SV40 large T protein. RB = Retinoblastoma protein.

to the full tumorigenic phenotype and that they each display very unique biological and biochemical activities. Thus, while the E6/E7 genes can cooperatively induce cellular immortalization, they do not induce tumorigenicity directly. In rare instances, however, prolonged passaging of HPV 18-immortalized cells has resulted in their conversion into tumorigenic cells. Also, the additional expression of the *ras* and *fos* genes in HPV-immortalized keratinocytes can also convert them to the tumorigenic state [34–36]. The E5 gene, while mitogenic in primary and immortalized cells, cannot immortalize cells. Despite the presence of several distinct transforming genes in the papillomaviruses, therefore, additional alterations in cellular gene expression are clearly needed for cells to acquire a malignant phenotype.

E7 Protein

Biochemical Properties

The human papillomavirus E7 proteins are acidic phosphoproteins consisting of approximately 100 amino acids which have transcriptional transactivation activity. Despite their predicted molecular weight of 11 kD, the E7 proteins exhibit an aberrant mobility on SDS-polyacrylamide gels with an apparent molecular weight of 14–21 kD [37]. While initial studies localized the E7 protein predominantly to the cytoplasm [38], subsequent immunofluorescence and cell fractionation assays have revealed that E7 also resides within the nuclear compartment [39]. Whether E7 nuclear translocation results from specific signal sequences or from its association with other cellular proteins is not known.

The E7 proteins share distinct amino acid homologies with two other DNA tumor virus oncoproteins: adenovirus E1A and the SV40 large T-antigen [40]. These homologous regions were first defined as conserved domains one and two (CDI and CDII) in adenovirus E1A proteins [41]. In the case of the HPV 16 E7 protein, the CDI-homologous region encompasses the first 15 N-terminal amino acids whereas the CDII-homologous region includes amino acids 16–38. At the C-terminal end of the CDII region lies a casein kinase II (CKII) recognition sequence which is the target of phosphorylation (fig. 1). In HPV 16 E7, serine residues 32 and 33 are phosphorylated [38, 42]. Although the C-terminal region of HPV 16 E7 (amino acids 38–98) has only minor homologies to the C-terminus of E1A, they both possess two zinc finger domains. The E7 proteins (as well as the E6 proteins discussed later) bind zinc through their Cys-X-X-Cys zinc finger motifs [43–45]. Binding of zinc ions by the Cys-X-X-Cys motifs is hypothesized to facilitate dimerization of two E7 proteins, and mutational analysis has also revealed the importance of this domain for protein stability [45, 46].

Association with Host Cellular Proteins

Retinoblastoma Protein (RB)

The E7 protein, comparable to the adenovirus E1A protein and the SV40 large T-antigen, binds to the retinoblastoma-susceptibility gene product (RB) [47–49]. The RB protein is a phosphoprotein which, in its underphosphorylated form, appears to negatively regulate entry into the S-phase of the cell cycle. Initiation of the S-phase is accompanied by the specific phosphorylation of the RB protein [for a review see ref. 50] via cyclin-dependent kinases (cdks) [for a review see ref. 51], and when cells enter mitosis the RB protein is dephosphorylated and returned to its growth-inhibitory form.

The E7 protein, similar to E1A and the large T-antigen, binds with highest affinity to the hypophosphorylated form of RB [52], thereby interfering with a critical growth regulatory step in the cell cycle. In the case of E1A, the CDI and CDII domains are important for RB binding [48, 53, 54]. For the E7 protein, however, only the CDII domain appears to mediate RB binding [46, 53, 55, 56]. However, there is a significant difference between E7 proteins of the various HPV types in interaction with the RB protein. Most interestingly, the E7 proteins of the HPVs frequently associated with the development of malignancy (high-risk HPVs) are most effective in binding RB, while the E7 proteins of HPVs infrequently associated with malignancy (low-risk HPVs) interact with RB less well [37, 47, 57–59].

RB-Related Proteins

The E7 protein also binds to the RB-related proteins p107 and p130 [53, 60, 61]. Interestingly, the binding of p107 to E7 is dependent upon CDII sequences which partially overlap the RB-binding site [61]. However, the functional consequence of these interactions has not been defined.

Cell Cycle Control Proteins

Cyclins and cdks have a critical role in regulating various phases of the cell cycle and could potentially represent another important target for the papillomavirus transforming proteins. In vitro assays have revealed that E7 protein expressed in yeast can bind to cyclin A and cdk2 present in cell extracts [62]. These results are supported by an additional study showing that the E7 protein binds to an unidentified histone H1 kinase [61] present in cervical carcinoma cell lines. If E7 can modulate the activity or localization of such kinases, it could profoundly affect the regulation of the cell cycle.

Transcription

Since the RB protein itself associates with human papillomavirus E7 proteins and several cell proteins, including the E2F transcription factor, cdk $p34^{cdc2}$, and phosphatase 1A, it is likely that E7 may modulate several of these interactions. Indeed, the binding of E7 to RB leads to the dissociation of an E2F/RB complex which appears to be an active inhibitor of transcription of genes possessing E2F-binding motifs [52, 63]. Binding of E7 would therefore result in the enhanced transcription of E2F-regulated genes, including several which are involved in cell division and DNA synthesis, such as c-*myc*, c-*myb*, cdc 2, DNA polymerase α, ribonucleotide reductase, and thymidylate synthase [64, 65].

Mutagenic analysis also supports an association between the ability of E7 to bind RB and gene transactivation [66, 67]. E7 mutants which are defective for RB binding (CDII mutants) are also completely defective for transactivating the adenovirus E2 promoter in rodent cells [66]. Interestingly, it is not possible to utilize the isolated E7 CDII domain (via the use of synthetic peptides) to functionally replace E7 [68] suggesting that CDII-binding interactions with RB are insufficient for transactivation. Although the CDII domain is a site of phosphorylation, the state of phosphorylation of E7 has only minor effects on the transactivating function of the E7 protein [66, 67, 69]. In contrast to E7 CDII mutants which are markedly reduced in their transactivating activity, only one CDI mutant (HPV16 E7 His2 to Asp) has been described [46] which decreases the transactivating function of E7. All other analyzed CDI mutants exhibit only minor alterations in transactivation function [66].

The C-terminus of E7 can be interchanged between low- and high-risk HPVs without interfering with its transactivational activity [58], and it is postulated that this domain physically interferes with the binding of E2F to RB.

Despite the different affinities of HPV E7 proteins (those derived from low-risk versus high-risk HPVs) for the RB protein, all E7 proteins exhibit similar transactivation of the adenovirus E2 promoter [58]. This correlates with the finding that an efficient transactivation function of the E7 protein is not required for its transforming abilities in immortalized rodent fibroblasts [67].

Transformation of Rodent and Human Cells

Immortalized Rodent Cells

The transforming properties of high-risk HPVs were first discovered with immortalized rodent fibroblast cells [24–27, 30]. When NIH 3T3 cells were transfected with dimeric HPV 16 genomes, they formed foci after prolonged culture. These foci were also capable of growing in soft agar and were tumorigenic in nude mice [27]. Genetic analysis indicated that it was the E7 gene in these assays which is responsible for focus formation [26–28, 30, 70–72]. In addition, it was discovered that there was a good correlation between the ability of E7 proteins to transform rodent cells and to bind RB protein [42, 57, 59]. Thus, only the E7 proteins of high-risk HPVs were effective at transforming rodent cells. However, it is possible to convert the E7 protein of low-risk HPVs to a transforming protein by mutating its CDII domain to a high-affinity RB-binding site [37]. The transforming potential of mutants at the CKII kinase site is described differently by various investigators. Phelps et al. [40] found no significant difference in the transforming ability of a deletion mutant, which lacks both phosphorylated serines in HPV 16 E7, whereas Barbosa et al. [42] and Watanabe et al. [46] concluded that mutating HPV 16 E7 serines 31 and 32 to arginine and proline, respectively, destroys the transforming function of the E7 protein on NIH 3T3 cells.

Despite the ability of E7 to mediate focus formation on rodent cells, it is only weakly mitogenic, at least when assayed on quiescent Swiss 3T3 cells [63]. The mitogenic activity of E7, however, can be augmented by the addition of growth factors such as insulin. Even though the CDI domain of E7 is not essential for RB binding, mutagenic analysis indicates that the CDI domain is essential for the transformation of immortalized rodent fibroblasts [56].

Primary Rodent Cells

Although the E7 gene is sufficient to transform immortalized rodent cells as outlined above, it is inactive on primary rodent cells. In order to achieve cell transformation of primary rodent cells, it is necessary to cotransfect the E7 gene with either an activated *ras, myc* [40, 73–75], or mutant p53 gene [76]. Thus, primary baby hamster kidney cells or embryonic fibroblasts can be simultaneously immortalized and transformed by E7 plus *ras*. The resultant cell lines grow in agarose and form tumors in nude mice. Similar to the situation with immortalized cells, the transforming activity of E7 correlated with its ability to form a complex with the RB protein [57, 58]. The ability of E7 to bind RB was found to be essential but not sufficient for primary cell transformation and, in order for E7 to cooperate with *ras*, the CDI domain needs to be intact [57, 58, 66]. Again, similar to findings with immortalized cell lines, the C-terminus of E7 is interchangeable between high-and low-risk HPVs without any loss of immortalizing or transforming activity of the chimeric E7 protein [58]. Concerning the influence of the phosphorylation state of the E7 protein on the transforming abilities on primary rodent cells, results have differed among investigators. Two studies report that deletion or mutation of the phosphorylated serines has no influence on the transforming potential of primary rodent cells [46, 66]. In a third study [69], at least a negatively charged amino acid is necessary at the position of the phosphorylated serines to obtain transformation by E7 in primary rodent cells.

Primary Human Keratinocytes

When expressed by an LTR promoter and transduced into cells by retroviral infection, the E7 gene of high-risk, but not low-risk, HPVs can independently immortalize human genital keratinocytes [22]. This immortalizing activity, however, can be synergistically increased by the coexpression of the E6 gene. In contrast, when expressed by the native HPV promoter, E7 appears inactive for genital keratinocyte immortalization unless coexpressed with the E6 gene [15, 18, 20, 21]. Presumably this difference reflects a low immortalization potential by isolated E7 which can be augmented by a strong promoter and highly efficient gene transfer. By itself, whether expressed by an LTR or its native HPV promoter, the E6 gene has no transforming activity on genital keratinocytes [20, 22]. When keratinocytes are immortalized by the isolated E7 gene, they exhibit an altered differentiation on raft cultures compared to normal primary genital keratinocytes, although this altered differentiation is not as pronounced as when both the E6 and E7 genes are expressed [17, 19]. In the raft culture system, coexpression of both E6 and E7 induces a cell phenotype which closely resembles cervical dysplasia. However, despite being immortalized, the E7-immortalized keratinocytes (as well as E6/E7-immortalized cells) are nontumorigenic in nude mice and do not grown in soft agar [13–15, 77].

Surprisingly, the immortalizing activity of E7 on primary human keratinocytes is independent of its ability to form complexes with RB [78]. This contrasts sharply with the findings of the rodent transformation system and indicates the importance of evaluating HPV gene function in the appropriate host cell.

While it is impossible to assess the function of HPV E6 and E7 genes for in vivo tumor formation, analysis of the cottontail rabbit papillomavirus (CRPV) indicates that the E7 gene is critical for tumor formation, although RB binding by E7 is dispensable [79, 80]. Thus, data from both in vitro and in vivo studies suggest that there is another critical target for E7 which is independent of RB.

In summary, it appears that the HPV E7 proteins are multifunctional and that the biological activities of E7 RB-binding mutants can vary with the specific cell being used as a target. It is anticipated that the interaction of E7 with unidentified cellular proteins (perhaps the cell cycle proteins) may represent a critical target which could potentially generate chromosomal aberrations such as aneuploidy.

E6 Protein

Biochemical Properties

The HPV E6 proteins are composed of approximately 150 amino acids and migrate on SDS-polyacrylamide gels with an apparent molecular weight of 16–18 kD. Similar to the E7 proeins, the E6 proteins contain multiple Cys-X-X-Cys motifs which apparently mediate the binding of zinc [43, 44] and represent a portion of an evolutionarily divergent 33-amino-acid unit [81]. Indeed, most of the E6 amino acid sequence appears to represent a duplication of the carboxyl-terminal region of E7. Mutagenic conversion of some of these conserved cysteines to glycine has demonstrated their importance in zinc binding and cellular transformation of rat 3Y1 cells [82]. Immunofluorescence microscopy has localized HPV 16 and 18 E6 proteins to the nuclear matrix as well as nonnuclear membrane fractions [44, 82]. The cysteine motifs appear to be important for this nuclear translocation since the E6 glycine mutants localize predominantly in the cytoplasm [82].

Associated Host Cell Proteins

The E1A and E1B proteins of adenovirus had been known previously to associate with the cellular RB and p53 tumor suppressor proteins, respectively [48, 83]. Based upon the findings that the E7 protein of HPV, similar to E1A, associated with the RB protein, subsequent studies also showed that the HPV E6 protein, similar to E1B, associated with p53 [83, 84]. The p53 protein was also a target for the SV40 large T-antigen [85, 86], indicating its common role in cellular transformation by DNA tumor viruses.

Although p53 is a common target for DNA-transforming proteins, the consequences of these interactions are quite different. Unlike the interactions between p53 and the large T-antigen or E1B, which stabilize the p53 protein, the E6/p53 interaction results in the degradation of the p53 protein [87]. This degradation is ubiquitin dependent and mediated through an interaction with a 100-kD cellular protein, designated E6-AP [88–90]. The degradation of p53 is mediated by the E6 proteins of the high-risk but not low-risk HPVs [84], suggesting an important role for this process in the development of malignancy. While some laboratories have shown that the lack of p53 degradation by low-risk HPVs is due to lack of E6/p53 binding [87], others have shown that the low-risk E6 proteins can bind, but not degrade, p53 [91]. In the latter study, mutagenic analysis defined the C-terminus of E6 as being important for p53 binding and the N-terminus for p53 degradation. All of the E6/p53 association/degradation studies have been performed using in vitro techniques. However, it is possible to demonstrate the interaction of E6/p53 in COS cells using an anti-p53 antibody [92].

E6 fusion proteins can also be used to mediate the degradation of proteins other than p53. For example, E7/E6 proteins containing the N-terminal half of the E7 protein were able to promote the degradation of the E7-associated RB protein [93].

The observation that cervical carcinoma cell lines positive for HPV DNA contain wild-type p53, while cervical lines negative for HPVs contain mutant p53 [94–96] argues that p53 degradation and mutation lead to the same functional consequences. However, other investigators have noted that HPV-negative cervical carcinomas may not necessarily contain p53 mutations [97, 98]. Several other lines of evidence further support the concept that a major function of E6 is to disrupt the normal tumor suppressor function of wild-type p53. (1) Human keratinocyte cell lines immortalized by HPV 16 or 18 show a 3- to 5-fold reduction in p53 half-life when compared to normal human keratinocytes, supporting the hypothesis that E6 mediates the degradation of p53 in vivo [92, 99]. (2) Mutant p53 can substitute for E6 in the immortalization of primary human keratinocytes even though mutant p53 does not possess E6-associated transactivation activity or transforming activity in 3T3 cells [100]. (3) Keratinocytes immortalized by HPV E6/E7 and cervical carcinoma cell lines harboring HPV 16 or 18 lose the G_1 arrest period following DNA damage [101]. This G_1 arrest is dependent upon wild-type p53 and allows damaged DNA to be repaired before reentering the cell cycle. This disruption of G_1 arrest was further shown to be due exclusively to the action of E6. When the isolated E6 gene was expressed in RKO carcinoma cells, they lost the ability to undergo G_1 arrest. (4) E6 interferes with the transactivation function of wild-type p53 as demonstrated by the inhibition of reporter gene transcription expressed by a p53-responsive promoter [102], or by the inhibition of the transcriptional transactivation by HeLa cell p53 [103]. Low-risk E6 proteins and E6 mutants defective for p53 binding are unable to inhibit p53-dependent transcription. (5) E6 abrogates p53-mediated repression of various reporters and this activity is unrelated to the transactivation function of E6 [92].

The absolute level of p53 in HPV-immortalized cells is not necessarily the critical determinant of cell transformation. For example, although keratinocytes immortalized by HPV generally contain reduced amounts of p53, they may also express levels of p53 equivalent to normal primary keratinocytes [94; Conrad Stöppler and Schlegel, unpubl. results]. In addition, while the half-life of p53 in HPV-immortalized keratinocytes is reduced, it is unaltered in E6-transformed 3T3 cells [100]. These findings support the hypothesis that transformation of 3T3 cells by E6 may either be p53-independent or that inactivation of murine p53 does not involve the same mechanism as described in keratinocytes. Finally, HeLa cells, which are known to express the HPV 18 E6/E7 genes, possess low levels of DNA-binding-competent and transcriptionally active p53 [103], suggesting that total inactivation of wild-type p53 may indeed be unnecessary for HPV-mediated oncogenesis.

Transformation of Rodent and Human Cells

In addition to its role in augmenting E7-mediated transformation of human genital keratinocytes, the HPV E6 gene also possesses its own independent transforming activities. Both the high-risk E6 and E7 genes (in the context of the HPV genome) are required for the efficient immortalization of newborn foreskin keratinocytes. However, when the host target cells are changed to primary human mammary epithelial cells, E6 can now generate immortal cell lines [104]. In fact, coexpression of E7 is not required. Evidently there are very significant differences in the regulatory pathways of genital squamous epithelium and mammary glandular epithelium such that the latter either do not require perturbation of RB function or already exhibit altered RB activity in order to undergo cellular immortalization. It is also apparent that this particular epithelium is particularly sensitive to interference with p53 regulation, a finding which is consistent with the unusually high occurrence of p53 mutations in breast cancer. Unexpectedly, the E6 genes of both high-risk and low-risk HPVs are able to facilitate cellular immortalization, although the high-risk E6 proteins appear to be the most efficient [105, 106]. Even the E6 protein from BPV 1 can inefficiently mediate immortalization. Despite the observation that the low-risk and BPV 1 E6 proteins do not directly promote the degradation of p53 in vitro, they do induce a reduction in cellular p53 in the immortalized cell lines. The mechanism by which this is achieved is not known.

When assayed in rodent cell systems, E6 exhibits similarities to the transforming activites of the E7 protein. Storey and Banks [107] showed that in baby rat kidney cells the HPV 16 E6 gene is able to induce immortalization in cooperation with an activated *ras,* while Liu et al. [108] have demonstrated that the HPV 16 E6 is able to fully transform baby mouse kidney cells in cooperation with activated *ras* [108]. The E6 protein, independent of *ras,* induces anchorage-independent growth in NIH 3T3 cells [21]. While mutant p53 was able to substitute for the HPV E6 protein in immortalization of primary human keratinocytes, mutant p53 was not able to substitute for E6 in transactivation assays or in transformation of 3T3 cells [100].

Transcriptional Transactivation by E6

The E6 proteins of HPV 6 and 16 can transactivate the adenovirus E2 promoter [21] and the HPV 16 E6 can additionally transactivate transcription from a series of heterologous viral promoters [109]. In one report, the HPV 18 E6 protein was able to stimulate transcription from its homologous upstream regulatory region [110]. The domain of E6 which effects transactivation is distinct from the domain required for p53 interaction. Mutants of HPV 16 E6 which are severely compromised for p53 binding/degradation retain the ability to transactivate the adenovirus E2 promoter. In addition, E6 proteins of the low-risk HPVs are as efficient at transactivation as those from high-risk viruses [91]. The transactivating activity of E6 therefore is not sufficient for the immortalization of genital keratinocytes and, extrapolating from the ability of a mutant p53 to substitute for E6 in immortalization of human keratinocytes but not in transactivation assays [100], it may even be unnecessary for immortalization. However, this certainly does not eliminate a potential contributory role in the natural pathogenesis of HPV infection.

E5 Protein

Biochemical Properties

The E5 proteins represent a unique class of transforming proteins. They are small (44–90 amino acids), extremely hydrophobic polypeptides which are present predominantly within intracellular membrane compartments, including the Golgi apparatus. Although being discussed along with E6 and E7, they cannot immortalize either rodent or human cells. Their main biological activity appears to be the mitogenic stimulation of cells via

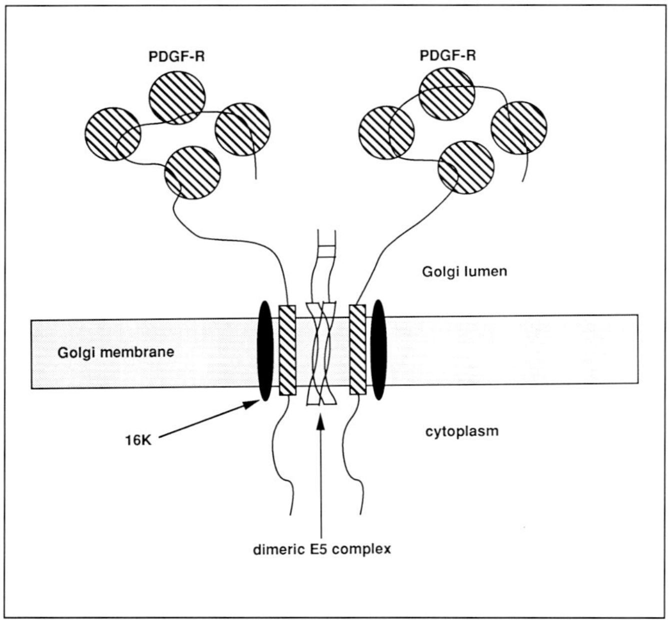

Fig. 2. Potential interactions of the BPV E5 oncoprotein with the PDGF receptor (R) and the 16K vacuolar pump protein. Due to its predominant localization in the Golgi and its ability to activate immature forms of the PDGF receptor, it is believed that E5 can mediate at least part of its mitogenic activity in Golgi and/or pre-Golgi membranes. E5, the PDGF receptor and 16K appear to exist in a ternary complex and cross-linking of this receptor complex by dimeric E5 represents a possible mechanism for activation of the PDGF receptor signal transduction pathway. Not shown is the described interaction of E5 with an adaptin-like molecule which is mediated via the carboxyl-terminus of E5.

growth factor receptor signal transduction pathways. The precise mechanism by which this activation of growth factor receptors occurs appears to depend on the particular E5 protein being studied.

BPV 1 E5 Protein

At the molecular level, the BPV 1 E5 oncoprotein is the best studied E5 protein. It is composed of 44 amino acids, the first 30 representing a hydrophobic membrane association domain. The final 14 carboxyl-terminal amino acids constitute a hydrophilic domain within which there are two highly conserved cysteine residues which mediate the dimerization and oligomerization of E5. Mutagenesis of either the hydrophobic or hydrophilic domain can abrograte the transforming activity of E5 [111–114]. By immunofluorescence and immunoelectronmicroscopy, E5 is found predominantly within Golgi membranes, although it can also be observed within intracellular vesicles and the plasma membrane [115].

In fibroblasts, the E5 protein associates with three different cellular proteins: the platelet-derived growth factor (PDGF) receptor [116], the 16K pore protein of the vacuolar proton pump [117, 118], and a 120-kD adaptin-like protein [119]. Interestingly, all three proteins can be found within the same cellular membrane compartments such as coated pits and endosomes, and both the adaptin-like and 16K proteins may have important roles in the internalization and degradation of activated growth factor receptors.

The principal growth stimulus effected by E5 in fibroblasts is the activation of PDGF receptors. Epithelial cells which lack the PDGF receptor are not transformable by E5 but can be rendered transformable by transfection with a cDNA encoding the PDGF receptor [120]. Interestingly, there is a remarkable specificity regarding which receptors E5 can activate. For example, E5 cannot activate the epidermal growth factor (EGF) receptor, the colony-stimulating factor (CSF)-1 receptor, or the PDGF α-receptor [121]. It can only activate the PDGF β-receptor. Whether this is the consequence of direct binding is not certain. However, E5 binds to the PDGF β-receptor but not the PDGF α-receptor or the EGF receptor [121, 122].

The above findings suggest that the simplest mechanism by which E5 can activate the PDGF β-receptor is by physical cross-linking (fig. 2). This would explain why monomeric forms of E5 are transformation defective [112]. While it has been suggested that E5 exhibits homology to PDGF and that it functions via mimicking the activity of this ligand (i.e. dimerization) [123], more recent data indicate that E5 binds and cross-links receptor complexes via transmembrane interactions [124] and not through interactions with the external ligand-binding domain of the receptor. Indeed, an extensive mutagenesis study of the E5 carboxyl-terminus was able to dissociate the PDGF homology of E5 from its transforming activity [114]. This same study also suggested that an acidic juxtamembrane amino acid (aspartic acid) in E5 was critical for its transforming activity and for mediating binding to a basic amino acid (lysine) in the external domain of the PDGF receptor. Interestingly, the corresponding amino acid in the binding-defective PDGF α-receptor is acidic and would not be predicted to mediate an ionic interaction. However, other studies have clearly shown that E5 can bind very efficiently to the PDGF receptor without the presence of aspartic acid and it is more likely that this acidic residue is involved in the molecular positioning of E5 within the plasma membrane. Precise transmembrane, helix/helix interactions have been postulated to regulate the binding of E5 to its cellular targets [124] and have recently been appreciated as critical determinants of membrane protein interactions [125].

It is also possible that E5 might activate or potentiate the activity of PDGF receptors in an indirect manner. For example, by interfering with the function of the vacuolar proton pump, E5 could decrease the degradation of internalized, activated receptors and thereby prolong the mitogenic stimulus. The Golgi might also represent another site of E5 action on the proton pump. Translocation of proteins through the Golgi to their correct destinations is apparently modulated by acidification of the trans-Golgi apparatus. By interfering with receptor transit through the Golgi, E5 might effect the accumulation of receptors in the Golgi which then become activated in a concentration-dependent manner. Consistent with such an hypothesis is the observation that overexpressed receptors are often 'spontaneously' activated due to increased physical interactions and transphosphorylations. However, it is not clear why the PDGF β-receptor would be the sole receptor sensitive to the aforementioned alterations. The EGF receptor is also internalized and degraded via acidic vesicles and processed in the Golgi in a manner very similar to the PDGF receptor. It would be necessary to speculate that the PDGF β-receptor possesses unique processing or biochemical properties which make it susceptible to the perturbation of the proton pump.

It also remains a possibility that the interaction of E5 with 16K might be indirect and due to E5/receptor interactions. Recently, 16K has been shown to be an independent component of receptor complexes [124; Andresson and Schlegel, unpubl. results] and the presence of 16K in such complexes may simply represent the fact that E5 interacts specifically with the PDGF receptor and coprecipitates receptor-associated 16K. However, association of E5 and 16K has been observed in cells lacking the PDGF receptor, suggesting direct binding [126].

Finally, E5 can also bind to a 120-kD adaptin-like protein which migh represent an independent mechanism for modulating the endocytosis of activated receptors and prolonging a receptor-initiated mitogenic response [119]. Adaptin proteins appear to play an important role in recognizing receptors in coated pits and governing their endocytosis and degradation. By preventing such recognition and degradation, E5 could maintain activated receptors for a longer period on the cell surface. This would be consistent with previous findings that E5 can cooperate with the EGF receptor in mediating the transformation of the fibroblasts and that E5 can apparently prolong the duration of receptors on the cell surface as well as decrease their rate of degradation [127, 128].

HPV E5 Proteins

Many, but not all, HPVs encode E5 proteins. E5 open reading frames are notably absent in the genomes of the epidermotropic papillomaviruses (e.g. HPV 5 and 8). However, the genital-associated low-risk HPVs (HPV-6, 11) and high-risk HPVs (e.g. HPV 16 and 18) all encode E5 proteins which are markedly hydrophobic and resemble the BPV 1 E5 protein when compared for hydrophilicity. However, the HPV E5 proteins are predicted to contain three separate transmembrane domains (rather than the one transmembrane domain of BPV 1) and exhibit little conservation of amino acid sequences [129]. In addition, the HPV E5 proteins have not been observed to dimerize or oligomerize as is the case with the BPV 1 E5 protein [130].

While clearly divergent in overall structure, the HPV 6 E5 protein does display several biochemical properties similar to the BPV 1 E5 protein. It is present in cell membranes and it binds to the 16K proton pump protein and the PDGF receptor [130, 131]. However, unlike the BPV 1 E5 protein, the HPV 6 E5 binds to additional tyrosine kinase growth factor receptors including the EGF and *erb*-b receptors [131]. Thus, the HPV 6 E5 protein appears to interact with a broader spectrum of growth factor receptors. In marked contrast, the HPV 16 E5 does not interact significantly with any of the above-mentioned growth factor receptors, suggesting that its transforming activity (see below) is not dependent upon direct binding to these receptors.The HPV 16 E5 protein does, however, interact with the 16K proton pump protein. Corresponding to this interaction, the HPV 16 E5 has been shown to interfere with the degradation of EGF receptors and to promote the recycling of these receptors back to the cell surface [132]. Such alterations are consistent with the HPV 16 E5 protein interfering with the activity of the vacuolar proton pump [McCance, unpubl. results].

Biological Activities

The HPV E5 proteins exhibit only weak transforming activity. For example, the HPV 16 E5 protein can induce the anchorage-independent growth of immortalized fibroblasts [132–134] but the frequency of cellular transformation is very low even after the selective isolation of E5-expressing cells. Interestingly, however, transformation by HPV 16 E5 is dependent upon coexpression of the EGF receptor and the presence of EGF, suggesting a specific cooperation with this receptor system.

Despite the ability of the HPV 16 E5 protein to induce anchorage-independent growth, it cannot reproducibly induce foci on immortalized fibroblasts (even in the presence of the EGF receptor and EGF), suggesting that dissimilar mechanisms operate for the BPV and HPV E5 proteins.

In summary, the E5 oncoprotein appears to induce cellular mitogenesis via signal transduction pathways. The mechanism by which E5 intercedes in and activates these pathways is unique but there is growing evidence that there may be related proteins expressed in retroviruses [135]. The newly observed specificity of transmembrane helix-helix interactions which mediate E5-target protein binding may also assist in defining the molecular mechanisms which regulate the interactions of integral transmembrane proteins.

References

1 zur Hausen H: Papillomaviruses in anogenital cancer as a model to understand the role of viruses in human cancer. Cancer Res 1989;49: 4677–4681.

2 zur Hausen H: Human papillomaviruses in the pathogenesis of human anogenital cancer. Virology 1991;184:9–13.

3 zur Hausen H: Human papillomaviruses and their possible role in squamous cell carcinomas. Curr Top Microbiol Immunol 1977;78:1–30.

4 Boshart M, Gissmann L, Ikenberg H, Kleinheinz A, Scheurlen W, zur Hausen H: A new type of papillomavirus DNA and its prevalence in genital cancer biopsies and in cell lines derived from cervical cancer. EMBO J 1984;3: 1151–1157.

5 Dürst M, Gissman L, Ikenberg H, zur Hausen H: A papillomavirus DNA from a cervical carcinoma and its prevalence in cancer biopsy samples from different geographic regions. Proc Natl Acad Sci USA 1983;80:3812–3815.

6 Gissmann L, Zur Hausen H: Partial characterization of viral DNA from human genital warts (condylomata acuminata). Int J Cancer 1980; 25:605–609.

7 Gissmann L, Diehl V, Schultz-Coulon HJ, zur Hausen H: Molecular cloning and characterization of human papillomavirus DNA derived from a laryngeal papilloma. J Virol 1982;4: 393–398.

8 Gissmann L, Boshart M, Dürst M, Ikenberg H, Wagner D, zur Hausen H: Presence of human papillomavirus (HPV) DNA in genital tumours. J Invest Dermatol 1984;83:265–288.

9 Ikenberg H, Gissmann L, Gross G, Gussendorf EI, zur Hausen H: Human papillomavirus type 16-related DNA in genital Bowen's disease and in Bowenoid papulosis. Int J Cancer 1983;32: 563–565.

10 Law MF, Lowy DR, Dvoretsky J, Howley PM: Mouse cells transformed by bovine papillomavirus contain only extrachromosomal viral DNA sequences. Proc Natl Acad Sci USA 1981;78:2727–2731.

11 Dvoretsky J, Shober R, Chattopadhyay SK, Lowy DR: A quantitative in vitro focus assay for bovine papillomavirus. Virology 1980;103: 369–375.

12 Black PH, Hartley JW, Rowe WP, Huebner RJ: Transformation of bovine tissue culture cells by bovine papillomavirus. Nature 1963;199: 1016–1018.

13 Dürst M, Dzarlieva-Petrusevska RT, Boukamp P, Fusenig NE, Gissmann L: Molecular and cytogenetic analysis of immortalized human keratinocytes obtained after transfection with human papillomavirus type 16 DNA. Oncogene 1987;1:251–256.

14 Pirisi L, Yasumoto S, Feller M, Doniger J, DiPaolo JA: Transformation of human fibroblasts and keratinocytes with human papillomavirus type 16 DNA. J Virol 1987;61:1061–1066.

15 Schlegel R, Phelps WC, Zhang YL, Barbosa M: Quantitative keratinocyte assay detects two biological activities of human papillomavirus DNA and identifies viral types associated with cervical carcinoma. EMBO J 1988;7:3181–3187.

16 Hawley-Nelson P, Vousden KH, Hubbert NL, Lowy DR, Schiller JT: HPV 16 E6 and E7 proteins cooperate to immortalize human foreskin keratinocytes. EMBO J 1989;8:3905–3910.

17 McCance DJ, Kopanj R, Fuchs E, Laimins LA: Human papillomavirus type 16 DNA alters human cell differentation in vitro. Proc Natl Acad Sci USA 1988;85:7169–7173.

18 Barbosa MS, Schlegel R: The E6 and E7 genes of HPV-18 are sufficient for inducing two-stage in vitro transformation of human keratinocytes. Oncogene 1989;4:1529–1532.

19 Hudson JB, Bedell MA, McCance DJ, Laimins LA: Immortalization and altered differentiation of human keratinocytes in vitro by the E6 and E7 open reading frames of human papillomavirus type 18. J Virol 1990;64:519–526.

20 Münger K, Phelps WC, Bubb V, Howley PH, Schlegel R: The E6 and E7 genes of the human papillomavirus type 16 together are necessary and sufficient for transformation of primary human keratinocytes. J Virol 1989;763:4417–4421.

21 Sedman SA, Barbosa MS, Vass WC, Hubbert NL, Haas JA, Lowy DR, Schiller JT: The full-length E6 protein of human papillomavirus type 16 has transforming and transactivating activities and cooperates with E7 to immortalize keratinocytes in culture. J Virol 1991;65: 4860–4866.

22 Halbert CL, Demers GW, Galloway DA: The E7 gene of human papillomavirus type 16 is sufficient for immortalization of human epithelial cells. J Virol 1991;65:473–478.

23 Halbert CL, Demers GW, Galloway DA: The E6 and E7 genes of human papillomavirus type 6 have weak immortalizing activity in human epithelial cells. J Virol 1992;66:2125–2134.

24 Kanda T, Watanabe S, Yoshike K: Human papillomavirus type 16 transformation of rat 3Y1 cells. Jpn J Cancer Res 1987;78:103–108.

25 Noda T, Yajima H, Ito Y: Progression of the phenotype of transformed cells after growth stimulation of cells by a human papillomavirus type 16 gene function. J Virol 1988;63:313–324.

26 Watanabe S, Yoshiike K: Transformation of rat 31Y cells by human papillomavirus type-18 DNA. Int J Cancer 1988;41:896–900.

27 Yasumoto S, Burkhardt A, Doniger J, DiPaolo J: Human papillomavirus type 16 DNA-induced malignant transformation of NIH 3T3 cells. J Virol 1986;57:572–577.

28 Bedell M, Jones K, Laimins L: The E6-E7 region of human papillomavirus type 18 is sufficient for transformation of NIH 3T3 and Rat-1 cells. J Virol 1987;61:3635–3640.

29 Chesters PM, Vousen KH, Edmonds C, McCance DJ: Analysis of human papillomavirus type 16 open reading frame E7 immortalizing function in rat embryo fibroblast cells. J Gen Virol 1990;71:449–453.

30 Kanda T, Watanabe S, Yoshiike K: Immortalization of primary rat cells by human papillomavirus type 16 subgenomic fragments controlled by the SV40 promoter. Virology 1988; 165:321–325.

31 Schwarz E, Freese UK, Gissmann L, Mayer W, Roggenbuck B, Stremlau A, zur Hausen H: Structure and transcription of human papillomavirus sequences in cervical carcinoma cells. Nature 1985;314:111–114.

32 Yee CL, Krishnan-Hewlett I, Baker CC, Schlegel R, Howley PM: Presence and expression of human papillomavirus sequences in human cervical carcinoma cell lines. Am J Pathol 1985;119:3261–3266.
33 von Knebel Doeberitz M, Rittmüller C, zur Hausen H, Dürst M: Inhibition of tumorigenicity of cervical cancer cells in nude mice by HPV E6-E7 antisense RNA. Int J Cancer 1992;51: 831–834.
34 DiPaolo JA, Woodworth CD, Popescu NC, Notario V, Doniger J: Induction of human cervical squamous cell carcinoma by sequential transfection with human papillomavirus 16 DNA and viral Harvey *ras*. Oncogene 1989;4: 395–399.
35 Dürst M, Gallahan D, Gilbert J, Rhim JS: Glucocorticoid-enhanced neoplastic transformation of human keratinocytes by human papillomavirus type 16 and an activated *ras* oncogene. Virology 1989;173:767–771.
36 Pei XF, Meck JM, Greenhalgh D, Schlegel R: Cotransfection of HPV-18 and v-*fos* DNA induces tumorigenicity of primary human keratinocytes. Virology 1993;196:855–860.
37 Sang BC, Barbosa MS: Single amino acid substitutions in 'low-risk' human papillomavirus (HPV) type 6 E7 protein enhance features characteristic of the 'high-risk' HPV E7 oncoproteins. Proc Natl Acad Sci USA 1992;89:8063–8067.
38 Smotkin D, Wettstein FO: The major human papillomavirus protein in cervical cancer is a cytoplasmic phosphoprotein. J Virol 1987;61: 1686–1689.
39 Sato H, Watanabe S, Furuno A, Yoshiike K: Human papillomavirus type 16 E7 protein expressed in *Escherichia coli* and monkey COS-1 cells: Immunofluorescence detection of the nuclear E7 protein. Virology 1989;170:311–315.
40 Phelps WC, Yee CL, Munger K, Howley PM: The human papillomavirus type 16 E7 gene encodes transactivation and transformation functions similar to those of adenovirus E1A. Cell 1988;53:539–547.
41 Moran E, Mathews MB: Multiple functional domains in the adenovirus E1A gene. Cell 1987;48:177–178.
42 Barbosa MS, Edmonds C, Fisher C, Schiller JT, Lowy DR, Vousden KH: The region of the HPV E7 oncoprotein homologous to adenovirus E1A and SV40 large T antigen contains separate domains for Rb binding and casein kinase II phosphorylation. EMBO J 1990;9:153–160.
43 Barbosa MS, Lowy DR, Schiller JT: Papillomavirus polypeptides E6 and E7 are zinc-binding proteins. J Virol 19089;63:1404–1407.
44 Grossman SR, Laimins LA: E6 protein of human papillomavirus type 18 binds zinc. Oncogene 1989;4:1089–1093.
45 McIntyre MC, Frattini MG, Grossman SR, Laimins LA: Human papillomavirus type 18 E7 protein requires intact Cys-X-X-Cys motifs for zinc binding, dimerization, and transformation but not for Rb binding. J Virol 1993;67: 3142–3150.
46 Watanabe S, Kanda T, Sato H, Furuno A, Yoshiike K: Mutational analysis of human papillomavirus type 16 E7 functions. J Virol 1990;64:207–214.
47 Dyson N, Howely PM, Münger K, Harlow E: The human papillomavirus-16 E7 oncoprotein is able to bind to the retinoblastoma gene product. Science 1989;243:934–937.
48 Whyte P, Buchkovich KJ, Horowitz JM, Friend SH, Raybuck M, Weinberg RA, Harlow E: Association between an oncogene and an anti-oncogene: The adenovirus E1A proteins bind to the retinoblastoma gene product. Nature 1988;33:124–129.
49 DeCaprio JA, Ludlow JW, Figge J, Shew JY, Huang CM, Lee WH, Marsilio E, Paucha E, Livingston DW: SV40 large tumor antigen forms a specific complex with the product of the retinoblastoma susceptibility gene. Cell 1988;54:275–283.
50 Goodrich DW, Lee WH: Molecular characterization of the retinoblastoma susceptibility gene. Biochim Biophys Acta 1993;1155:43–61.
51 Motokura T, Arnold A: Cyclins and oncogenesis. Biochim Biophys Acta 1993;1155:63–78.
52 Imai Y, Matsushima Y, Sugimura T, Terada M: Purification and characterization of human papillomavirus type 16 E7 protein with preferential binding capacity to the underphosphorylated form of retinoblastoma gene product. J Virol 1991;65:4966–4972.
53 Dyson N, Guida P, Münger K, Harlow E: Homologous sequences in adenovirus E1A and human papillomavirus E7 proteins mediate interaction with the same set of cellular proteins. J Virol 1992;66:6893–6902.
54 Dyson N, Guida P, McCall C, Harlow E: Adenovirus E1A makes two distinct contacts with te retinoblastoma protein. J Virol 1992;66: 4606–4611.
55 Stirdivant SM, Huber HE, Patrick DR, Defeo JD, McAvoy EM, Garsky VM, Oliff A, Heimbrook DC: Human papillomavirus type 16 E7 protein inhibits DNA binding by the retinoblastoma gene product. Mol Cell Biol 1992;12: 1905–1914.
56 Banks L, Edmonds C, Vousden KH: Ability of the HPV 16 E7 protein to bind RB and induce DNA synthesis is not sufficient for efficient transforming activity in NIH3T3 cells. Oncogene 1990;5:1383–1389.
57 Heck DV, Yee CL, Howley PM, Münger K: Efficiency of binding the retinoblastoma protein correlates with the transforming capacity of the E7 oncoproteins of the human papillomaviruses. Proc Natl Acad Sci USA 1992;89: 4442–4446.
58 Münger K, Yee CL, Phelps WC, Pietenpol JA, Moses HL, Howely PM: Biochemical and biological differences between E7 oncoproteins of the high- and low-risk human papillomavirus types are determined by amino-terminal sequences. J Virol 1991;65:3943–3948.
59 Gage JR, Meyers C, Wettstein FO: The E7 proteins of the nononcogenic human papillomavirus type 6b (HPV-6b) and of the oncogenic HPV-16 differ in retinoblastoma protein binding and other properties. J Virol 1990;64:723–730.
60 Davies RC, Vousden KH: Functional analysis of human papillomavirus type 16 E7 by complementation with adenovirus E1A mutants. J Gen Virol 1992;73:2135–2139.
61 Davies R, Hicks R, Crook T, Morris J, Vousden K: Human papillomavirus type 16 E7 associates with a histone H1 kinase and with p107 through sequences necessary for transformation. J Virol 1993;67:2521–2528.
62 Tommasino M, Adamczewski JP, Carlotti F, Barth CF, Manetti R, Contorni M, Cavalieri F, Hunt T, Crawford L: HPV 16 E7 protein associates with the protein kinase p33CDK2 and cyclin A. Oncogene 1993;8:195–202.
63 Morris JDH, Crook T, Bandara LR, Davies R, LaThangue NB, Vousden KH: Human papillomavirus type 16 E7 regulates E2F and contributes to mitogenic signalling. Oncogene 1993;8: 893–898.
64 Mudryi M, Hiebert SW, Nevins JR: A role for the adenovirus inducible E2F transcription factor in a proliferation dependent signal transduction pathway. EMBO J 1990;9:2179–2184.
65 Dalton S: Cell cycle regulation of the human cdc2 gene. EMBO J 1992;11:1797–1804.
66 Phelps WC, Münger K, Yee CL, Barnes JA, Howley PM: Structure-function analysis of the human papillomavirus type 16 E7 oncoprotein. J Virol 1992;66:2418–2427.
67 Edmonds C, Vousden KH: A point mutational analysis of human papillomavirus type 16 E7 protein. J Virol 1989;63:2650–2656.
68 Huang PS, Patrick DR, Edwards G, Goodhart PJ; Huber HE, Miles L, Garsky VM, Oliff A, Heimbrook DC: Protein domains governing interactions between E2F, the retinoblastoma gene product, and human papillomavirus type 16 E7 protein. Mol Cell Biol 1993;13:953–960.
69 Firzlaff JM, Luscher B, Eisenman RN: Negative charge at the casein kinase II phosphorylation site is important for transformation but not for Rb protein binding by the E7 protein of human papillomavirus type 16. Proc Natl Acad Sci USA 1991;88:5187–5191.
70 Vousden KH, Doniger J, DiPaolo JA, Lowy DR: The E7 open reading frame of human papillomavirus type 16 encodes a transforming gene. Oncogene Res 1988;3:167–175.
71 Tanaka A, Noda T, Yajima H, Hatanaka M, Ito Y: Identification of a transforming gene of human papillomavirus type 16. J Virol 1989; 63:1465–1469.
72 Bedell MA, Jones KH, Grossman SR, Laimins LA: Identification of human papillomavirus type 18 transforming genes in immortalized and primary cells. J Virol 1989;63:1247–1255.
73 Chesters PM, McCance DJ: Human papillomavirus types 6 and 16 in cooperation with Ha-*ras* transform secondary rat embryo fibroblasts. J Gen Virol 1989;70:353–365.

74 Peacock JW, Matlashewski GJ, Benchimol S: Synergism between pairs of immortalizing genes in transformation assays of rat embryo fibroblasts. Oncogene 1990;5:1769–1779.
75 Storey A, Pim D, Murray A, Osborn K, Banks L, Crawford L: Comparison of the in vitro transforming activities of human papillomavirus types. EMBO J 1988;7:1815–1820.
76 Crook T, Fisher C, Vousden KH: Modulation of immortalizing properties of human papillomavirus type 16 E7 by p53 expression. J Virol 1991;65:505–510.
77 Woodworth CD, Bowden PE, Doniger J, Pirisi L, Barnes W, Lancaster WD, DiPaolo JA: Characterization of normal human exocervical epithelial cells immortalized in vitro by papillomavirus types 16 and 18 DNA. Cancer Res 1988;48:4620–4628.
78 Jewers RJ, Hildebrandt P, Ludlow JW, Kell B, McCance DJ: Regions of human papillomavirus type 16 E7 oncoprotein required for immortalization of human keratinocytes. J Virol 1992;66:1329–1335.
79 Brandsma JL, Yang ZH, Barthold SW, Johnson EA: Use of a rapid, efficient inoculation method to induce papillomas by cottontail rabbit papillomavirus DNA shows that the E7 gene is required. Proc Natl Acad Sci USA 1991; 88:4816–4820.
80 Defeo-Jones D, Vuocolo GA, Haskell KM, Hanobik MG, Kiefer DM, McAvoy EM, Ivey HM, Brandsma JL, Oliff A, Jones RE: Papillomavirus E7 protein binding to the retinoblastoma protein is not required for viral induction of warts. J Virol 1993;67:716–725.
81 Cole ST, Danos O: Nucleotide sequence and comparative analysis of the human papillomavirus type 18 genome: Phylogeny of papillomaviruses and repeated structure of the E6 and E7 gene products. J Mol Biol 1987;193:599–608.
82 Kanda T, Watanabe S, Zanma S, Sato H, Furuno A, Yoshiike K: Human papillomavirus type 16 E6 proteins with glycine substitution for cysteine in the metal-binding motif. Virology 1991;185:536–543.
83 Sarnow P, Ho YS, Williams J, Levine AJ: Adenovirus E1b-58kd tumor antigen and SV40 large tumor antigen are physically associated with the same 54 kd cellular protein in transformed cells. Cell 1982;28:387–394.
84 Werness BA, Levine AJ, Howley PM: Association of human papillomavirus types 16 and 18 E6 proteins with p53. Science 1990;248:76–79.
85 Lane D, Crawford L: T antigen is bound to a host protein in SV40-transformed cells. Nature 1979;278:261–263.
86 Linzer DIH, Levine AJ: Characterization of a 54K dalton cellular SV40 tumor antigen present in SV40-transformed cells and uninfected embryonal carcinoma cells. Cell 1979; 17:43–52.
87 Scheffner M, Werness BA, Huibregtse JM, Levine AJ, Howley PM: The E6 oncoprotein encoded by human papillomavirus types 16 and 18 promotes the degradation of p53. Cell 1990; 63:1129–1136.
88 Huibregtse JM; Scheffner M, Howley PM: A cellular protein mediates association of p53 with the E6 oncoprotein of human papillomavirus types 16 or 18. EMBO J 1991;10:4129–4135.
89 Huibregtse JM; Scheffner M, Howley PM: Localization of the E6-AP regions that direct human papillomavirus E6 binding, association with p53, and ubiquitination of associated proteins. Mol Cell Biol 1993;13:4918–4927.
90 Huibregtse JM; Scheffner M, Howley PM: Cloning and expression of the cDNA for E6-Ap, a protein that mediates the interaction of the human papillomavirus E6 oncoprotein with p53. Mol Cell Biol 1993;13:755–784.
91 Crook T, Tidy JA, Vousden KH: Degradation of p53 can be targeted by HPV E6 sequences distinct from those required for p53 binding and trans-activation. Cell 1991;67:547–556.
92 Lechner MS, Mack DH, Finicle AB, Crook T, Vousden KH, Laimins LA: Human papillomavirus E6 proteins bind p53 in vivo and abrogate p53-mediated repression of transcription. EMBO J 1992;11:3045–3052.
93 Scheffner M, Münger K, Huibregtse JM, Howley PM: Targeted degradation of the retinoblastoma protein by human papillomavirus E7-E6 fusion proteins. EMBOJ 1992;11: 2425–2431.
94 Scheffner M, Münger K, Byrne JC, Howley PM: The state of the p53 and retinoblastoma genes in human cervical carcinoma cell lines. Proc Natl Acad Sci USA 1991;88:5523–5527.
95 Wrede D, Tidy JA, Crook T, Lane D, Vousden KH: Expression of RB and p53 proteins in HPV-positive and HPV-negative cervical carcinoma cell lines. Mol Carcinog 1991;4: 171–175.
96 Crook T, Wrede D, Tidy JA, Mason WP, Evans DJ, Vousden KH: Clonal p53 mutation in primary cervical cancer: Assocation with human-papillomavirus-negative tumours. Lancet 1992;339:1070–1073.
97 Borresen AL, Helland A, Nesland J, Holm R, Trope C, Kaern J: Papillomaviruses, p53, and cervical cancer (letter). Lancet 1992;339: 1350–1351.
98 Busby-Earle RMC, Steel CM, Williams ARW, Cohen B, Bird CC: Papillomaviruses, p53, and cervical cancer (letter). Lancet 1992; 339:1350.
99 Hubbert NL, Sedman SA, Schiller JT: Human papillomavirus type 16 increases the degradation rate of p53 in human keratinocytes. J Virol 1992;66:6237–6241.
100 Sedman SA, Hubbert NL, Vass WC, Lowy DR, Schiller JT: Mutant p53 can substitute for human papillomavirus type 16 E6 in immortalization of human keratinocytes but does not have E6-associated trans-activation or transforming activity. J Virol 1992;66: 4201–4208.
101 Kessis TD, Slebos RJ, Nelson WG, Kastan MB, Plunkett BS, Han SM, Lorincz AT, Hedrick L, Cho KR: Human papillomavirus 16 E6 expression disrupts the p53-mediated cellular response to DNA damage. Proc Natl Acad Sci USA 1993;90:3988–3992.
102 Mietz JA, Unger T, Huibregtse JM; Howley PM: The transcriptional transactivation function of wild-type p53 is inhibited by SV40 large T-antigen and by HPV-16 E6 oncoprotein. EMBO J 1992;11:5013–5020.
103 Hoppe-Seyler F, Butz K: Repression of endogenous p53 transactivation function in HeLa cervical carcinoma cells by human papillomavirus type 16 E6, human mdm-2, and mutant p53. J Virol 1993;67:3111–3117.
104 Band V, Zajchowski D, Kulesa V, Sager R: Human papillomavirus DNAs immortalize normal human mammary epithelial cells and reduce their growth factor requirements. Proc Natl Acad Sci USA 1990;87:463–467.
105 Band V, DeCaprio JA, Delmolino L, Kulesa V, Sager R: Loss of p53 protein in human papillomavirus type 16 E6-immortalized human mammary epithelial cells. J Virol 1991; 65:6671–6676.
106 Band V, Dalal S, Delmolino L, Androphy EJ: Enhanced degradation of p53 protein in HPV-6 and BPV-1 immortalized human mammary epithelial cells. EMBO J 1993;12: 1847–1852.
107 Storey A, Banks L: Human papillomavirus type 16 E6 gene cooperates with EJ-*ras* to immortalize primary mouse cells. Oncogene 1993;8:919–924.
108 Liu Z, GhaiJ, Ostrow RS, McGlennen RC, Faras AJ: The E6 gene of human papillomavirus type 16 is sufficient for transformation of baby rat kidney cells in cotransfection with activated Ha-*ras.* Virology 1993;201:388–396.
109 Desaintes C, Hallez S, Alphen PV, Burney A: Transcriptional activation of several heterologous promoters by the E6 protein of human papillomavirus type 16. J Virol 1992;66:325–333.
110 Gius D, Grossman S, Bedell MA, Laimins LA: Inducible and constitutive enhancer domains in the noncoding region of human papillomavirus type 18. J Virol 1988;62:665–672.
111 Kulke R, Horwitz BH, Zibello T, DiMaio D: The central hydrophobic domain of the bovine papillomavirus E5 transforming protein can be functionally replaced by many hydrophobic amino acid sequences containing a glutamine. J Virol 1992;66:505–511.
112 Horwitz BH, Burkhardt A, Schlegel R, DiMaio D: 44-amino acid E5 transforming protein of bovine papillomavirus requires a hydrophobic core and specific carboxyl-terminal amino acids. Mol Cell Biol 1988;8:4071–4078.
113 Sparkowski J, Anders J, Schlegel R: Mutation of the bovine papillomaviurs E5 oncoprotein at amino acid 17 generates both high and low transforming variants. J Virol, in press.

114 Meyer A, Xu YF, Webster MK, Smith AE, Donoghue DJ: Cellular transformation by a transmembrane peptide: Structural requirements for the bovine papillomavirus E5 protein. Proc Natl Acad Sci USA 1994;91:4634–4638.
115 Burkhardt A, Willingham M, Gay C, Jeang KT, Schlegel R: The E5 oncoprotein of bovine papillomavirus is oriented asymmetrically in Golgi and plasma membranes. Virology 1989;170:2381–2385.
116 Petti L, DiMaio D: Stable association between the bovine papillomavirus E5 transforming protein and activated platelet-derived growth factor receptor in transformed mouse cells. Proc Natl Acad Sci USA 1992; 89:6736–6740.
117 Goldstein DJ, Schlegel R: The E5 oncoprotein of bovine papillomavirus binds to a 16 kd cellular protein. EMBO J 1990;9:137–146.
118 Goldstein DJ, Finbow ME, Andresson T, McLean P, Smith K, Bubb V, Schlegel R: The bovine papillomavirus E5 oncoprotein binds to the 16 kilodalton component of vacuolar H^+-ATPases. Nature 1991;352:347–349.
119 Cohen BD, Lowy DR, Schiller JT: The conserved C-terminal domain of the bovine papillomavirus E5 oncoprotein can associate with an alpha-adaptin-like molecule: A possible link between growth factor receptors and viral transformation. Mol Cell Biol 1993;13: 6462–6468.
120 Nilson LA, DiMaio D: Platelet-derived growth factor receptor can mediate tumorigenic transformation by the bovine papillomavirus E5 oncoprotein. Mol Cell Biol 1993; 13:4137–4145.
121 Goldstein DJ, Li W, Wang LM, Heidaran M, Aaronson S, Shinn R, Schlegel R, Pierce JH: The bovine papillomavirus type 1 E5 transforming protein specifically binds and activates the β-type receptor for the platelet-derived growth factor but not other related tyrosine kinase-containing receptors to induce cellular transformation. J Virol 1993;68: 4432–4441.
122 Petti L, DiMaio D: Specific interaction between the bovine papillomavirus E5 transforming protein and the β receptor for platelet-derived growth factor in stably transformed and acutely transfected cells. J Virol 1993;68:3582–3592.
123 Petti L, Nilson L, DiMaio D: Activation of the platelet-derived growth factor receptor by the bovine papillomavirus E5 transforming protein. EMBO J 1991;10:845–855.
124 Goldstein DJ, Andresson T, Sparkowski JJ, Schlegel R: The BPV-1 E5 protein, the 16 kD membrane pore-forming protein, and the PDGF receptor exist in a complex which is dependent upon hydrophobic transmembrane interactions. EMBO J 1992;11:4851–4859.
125 Lemmon MA, Engleman DM: Specificity and promiscuity in membrane helix interactions. Q Rev Biophys, in press.
126 Toyama R, Goldstein DJ, Schlegel R, Dhar R: A genomic sequence of the *Schizosaccharomyces pombe* 16 kDa vacuolar H-ATPase. Yeast 1991;7:989–991.
127 Martin P, Vass WC, Schiller JT, Lowy DR, Velu TJ: The bovine papillomavirus E5 transforming protein can stimulate the transforming activity of EGF and CSF-1 receptors. Cell 1989;59:21–32.
128 Waters CM, Overholser KA, Sorkin A, Carpenter G: Analysis of the influence of the E5 transforming protein on kinetic parameters of epidermal growth factor binding and metabolism. J Cell Physiol 1992;152:253–263.
129 Bubb V, McCance DJ, Schlegel R: DNA sequence of the HPV-16 E5 ORF and the structural conservation of its encoded protein. Virology 1988;163:243–246.
130 Conrad M, Bubb VJ, Schlegel R: The HPV-6 and HPV-16 E5 proteins are membrane-associated proteins which associate with the 16 kD pore-forming protein. J Virol 1993;67: 6170–6178.
131 Conrad M, Goldstein D, Andresson T, Schlegel R: The E5 protein of HPV-6, but not HPV-16, associates efficiently with cellular growth factor receptors. Virology 1994;200: 796–800.
132 Straight SW, Hinkle PM, Jewers RJ, McCance DJ: The E5 oncoprotein of human papillomavirus type 16 transforms fibroblasts and effects the downregulation of the epidermal growth factor receptor in keratinocytes. J Virol 1993;67:4521–4532.
133 Pim D, Collins M, Banks L: Human papillomavirus type 16 E5 gene stimulates the transforming activity of the epidermal growth factor receptor. Oncogene 1992;7:27–32.
134 Leechanachai P, Banks L, Moreau F, Matlashewski G: The E5 gene from human papillomavirus type 16 is an oncogene which enhances growth factor-mediated signal transduction to the nucleus. Oncogene 1992;7:19–25.
135 Franchini G, Mulloy JC, Koralnik IJ, Lo Monico A, Sparkowski JJ, Andresson T, Goldstein DJ, Schlegel R: The human T-cell leukemia/lymphotropic virus type I p12I protein cooperates with the E5 oncoprotein of bovine papillomavirus in cell transformation and binds the 16-kilodalton subunit of the vacuolar H-ATPase. J Virol 1993;67:7701–7704.

Intervirology 1994;37:180–188

Annette Altmann
Ingrid Jochmus
Frank Rösl

Deutsches Krebsforschungszentrum, Forschungsschwerpunkt Angewandte Tumorvirologie, Heidelberg, Deutschland

Intra- and Extracellular Control Mechanisms of Human Papillomavirus Infection

Key Words
Control mechanisms
Tumor suppression
Transcription factors
Cytokines
T lymphocytes
Macrophages
Accessory molecules
Natural killer cells

Introduction

Particular types of human papillomaviruses (such as HPV 16/18) are etiologically involved in the development of cervical cancer [1, 2]. Although the expression of the oncogenic reading frames E6 and E7 from high-risk HPV types is required for immortalization of human keratinocytes or fibroblasts in vitro, the cells are not tumorigenic upon heterotransplantation into nude mice [3–6]. Only the long-term propagation of these cultures in vitro [7] or the transfection of an additional activated oncogene leads to the generation of tumorigenic cells [8, 9]. In agreement with the multistep concept of carcinogenesis [10], these findings support epidemiological data [11] which also suggest that virus infection per se is necessary but not sufficient to malignantly transform human cells [2].

What are the regulatory mechanisms which usually control viral persistence during latency? In addition to an *extra*cellular immunological network which normally protects the host from virus infection [12], it has been postulated that the viral regulatory region which is involved in regulating the expression rate of the E6/E7 oncoproteins is a target of *intra*cellular control mechanisms negatively interfering with HPV transcription. Consequently, the monoclonal derivation of malignant cells can be regarded as loss of an intracellular surveillance mechanism as well as escape from immunological mechanisms controlling persistent HPV expression [13, 14].

Intracellular Control Mechanisms

Considering the intracellular pathway, one way to disturb the normal cell cycle after viral infection is the interaction of the oncogenic proteins E6/E7 with potential tumor suppressor genes such as p53 [15] or the retinoblastoma protein (pRB) [16, 17]. In the case of p53, interaction with E6 from high-risk HPV results in a ubiquitin-directed degradation of p53 while the corresponding protein of HPV 6 or 11 does not possess such a property [18]. Another important step in disturbing cell cycle control is the functional interference of HPV E7 with complex formation of pRB and the transcription factor E2F [19]. E7 is capable of destroying the interaction between pRB and E2F leading to a dissociation of E2F transcription factor [20] which in turn can transactivate E2F-dependent cellular host genes such as B-*myb* [21].

Annette Altmann
Deutsches Krebsforschungszentrum
Forschungsschwerpunkt Angewandte Tumorvirologie
Im Neuenheimer Feld 242
D–69120 Heidelberg (Germany)

0300–5526/94/0374–0180
$8.00/0

On the other hand, despite these properties of the viral oncoproteins, transfected primary recipient cells only become immortalized but are nonmalignant in vivo [3–6]. Examination of HPV expression in such cells after inoculation into nude mice by in situ hybridization reveals a selective down-regulation of viral transcription in nontumorigenic cells. In contrast, HPV-16- or HPV-18-positive cervical carcinoma cells form rapidly growing tumors with a high mitotic index showing viral gene expression equally distributed over the whole tissue section [22, 23]. This observation supports the notion that viral transcription is differently regulated in nontumorigenic and tumorigenic cells in immunocompromised animals [13].

Another interesting aspect in understanding the different progression steps of HPV-linked carcinogenesis is the finding that constitutively expressed viral oncogenes [24] such as the SV40 large T antigen [25], E1a protein of adenovirus [26] and also E7 of HPV 16 [27] are capable of inducing chromosomal changes of the host karyotype, finally leading to polyploidy and chromosomal segregation.

In recent years it has become evident that the phenotype of tumorigenicity in cervical carcinoma cells can indeed be correlated with the loss of genetic function. These data are mainly based on cell fusion experiments between HPV-positive tumorigenic cells and normal human keratinocytes or human fibroblasts [28]. Fusion of cervical carcinoma cell lines with either normal cells or microcells containing individual chromosomes results in suppression of the tumorigenic phenotype [29, 30]. As shown by restriction enzyme fragment length polymorphism in combination with karyotyping studies, chromosome 11 seems to be involved in the suppression effect [31].

Initial hints for a differential regulation of HPV 18 gene expression in malignant versus nonmalignant cells were provided by studies with somatic cell hybrids after treatment with substances involved in cell differentiation and DNA methylation [32]. It has been shown that 5-azacytidine can selectively suppress HPV 18 transcription in nontumorigenic cells at the level of transcription initiation which is accompanied by a cessation of cellular growth. Tumorigenic segregants as well as the parental HeLa cells are refractory to 5-azacytidine treatment [33]. Since the expression of HPV 16/18 is not only responsible for initiation but also for the maintenance of the transformed phenotype [34, 35], it is tempting to speculate that those genes which negatively regulate viral transcription and, consequently, the proliferation rate of the cells could be down-regulated by DNA-hypermethylation. Indeed, it has been shown recently that genes which are not compatible with cell differentiation under in vitro conditions are epigenetically modified via de novo methylation [36].

To detect negative regulation of the viral control region without the accompanying reduction of the cell proliferation rate, a reporter construct directed by the HPV 18 regulatory region was introduced in heterologous HPV-16-positive cervical carcinoma cells. The reporter gene which is constitutively expressed in malignant cells becomes completely extinguished after fusion with nonmalignant human keratinocytes. Both cycloheximide treatment and intracellular competition studies demonstrate that the extinction is mediated by repressor molecules derived from the nonmalignant fusion partner [37].

The importance of chromosome 11 in the regulation of high-risk HPV in infected cells is also supported by another set of experiments using human fibroblasts which harbor a deletion in the short arm of chromosome 11 [del(11p15)] [38]. In contrast to human fibroblasts with a normal chromosome set, del(11) cells allow high transcriptional activity of the HPV 16 regulatory region. This property seems to be mediated by a differential regulation of the phosphatase 2A subunits since inhibition of this enzyme by okadaic acid also provides activation of the HPV 16 regulatory region in normal human fibroblasts [39].

Which kind of transcription factors can potentially affect the regulatory region of human pathogenic papillomaviruses?

It has been demonstrated that both Oct-1 [40] and YY1 [41] are capable of suppressing the regulatory region of HPV 18 under transient transfection conditions. While the physiological role of Oct-1 in HPV regulation remains to be elucidated, YY1 seems to be involved in negative regulation of persistent HPV infection. Episomal HPV 16 DNA molecules detected in biopsies of malignant lesions obviously escape from cellular repression through mutations or deletions in their YY1 consensus sequences, thereby allowing higher transcriptional activity of the corresponding regulatory control region [42].

Furthermore, retinoic acid, which is an important regulator of epithelial cell growth and cell differentiation, has an inhibitory effect on HPV 18 transcription in HeLa cells. The retinoic acid receptor gene (RAR-β) seems to be differently regulated in somatic cell hybrids since the gene is only inducible in nontumorigenic HeLa/normal cell hybrids [43]. This finding is of particular interest especially in the context of a tumor suppressive effect of RAR-β in human epidermoid lung cancer cells [44].

In order to understand the various control levels under in vivo conditions, detailed in situ hybridization studies in combination with a histochemical evaluation of the naturally occurring and experimentally induced tumors are necessary.

Whether certain transcription factors also play a role during the different stages of cervical intraepithelial neoplasia (CIN) and whether malignant progression can be correlated with changes in viral expression and microenvironmental modifications of the tissue during tumor progression will be tested.

Of particular interest are, of course, those transcription factors which can be transcriptionally modulated by different cytokines. For example, overexpression of NF-IL-6 which has recognition sites in the viral control region can negatively interfere with HPV 16 expression in CaSki cells [45], but the effect of interleukin (IL)-6 on the expression of HPV is not clear.

How do cytokines interact with viral gene expression and how do they reach their target cells? As demonstrated recently, transcription of the JE gene encoding the monocyte-chemoattractant protein (MCP-1) [46] is not detectable in cervical carcinoma cells but the gene becomes reexpressed in nonmalignant somatic cell hybrids [47]. Hence, MCP-1 seems to be an attractive candidate involved in communication with the immunological network, since the protein not only attracts mononuclear cells to infiltrate into the tissue, but also activates them to secrete a whole variety of different cytokines [12] whose properties will be discussed below.

Extracellular Control Mechanisms

Cytokines

Cytokines Produced by Keratinocytes

Recent studies have shown that different lymphokines are secreted by cultured human keratinocytes [48, 49]. One of these lymphokines is IL-1 which induces a wide variety of inflammatory and immune functions, such as production of IL-2 by T helper cells. Keratinocytes themselves are stimulated by IL-1 to proliferate [50] and to synthesize and secrete other lymphokines including IL-6 [51], IL-8, granulocyte-macrophage-colony-stimulating factor (GM-CSF), and tumor necrosis factor (TNF-α) [52]. Each of these molecules promotes inflammation or stimulates host immunity by specific mechanisms. Cervical keratinocytes immortalized by HPV 16 or 18 show a significantly reduced production of these cytokines when compared to normal cells [52]. This seems to be due to secondary cellular changes following HPV expression and immortalization. In contrast to these results, Malejczyk et al. [53] demonstrated that the HPV-16-harboring and -expressing keratinocyte line SK-v constitutively releases IL-6 and TNF-α. TNF-α seems to exert an autocrine antiproliferative effect [54], which could be due to repression of HPV 16 E6 and E7 transcription [55]. Only highly tumorigenic sublines of the parental SK-v cells escape from this TNF-α-mediated growth limitation, probably by down-regulation of TNF-α receptor expression [56]. Additionally, soluble TNF-α receptor with inhibitory properties was found in SK-v supernatants [57]. On the presumption that constitutive release of cytokines is part of the immunoprotection, TNF-α and IL-6 production by HPV-16-positive cells might represent a self-control mechanism indirectly preventing the growth and spread of virus-induced neoplasia. Tumorigenic progression could result from switching off this autocrine regulation mechanism. On the other hand, the decreased secretion of cytokines by HPV-infected keratinocytes as observed by Woodworth et al. [52] might favor virus persistence.

Recently, Majewski et al. [58] observed an increased expression of the transforming growth factor (TGF)-β_1 and of TNF-α transcripts in HPV-5- or HPV-8-induced skin lesions of patients suffering from epidermodysplasia verruciformis (EV) but not in HPV-unrelated premalignant or malignant skin lesions. TGF-β_1 functionally resembles TNF-α in several ways. For instance, it also acts as a negative autocrine regulator of epithelial cell growth and differentiation. In the case of TGF-β_1, down-regulation of c-*myc* expression mediated by pRB or related cellular proteins is postulated as a mechanism for this effect [59]. According to Pietenpol et al. [59], proliferation of HPV-16- or HPV-18-positive keratinocytes is not inhibited by TGF-β_1 which seems to be due to an abrogation of TGF-β-induced c-*myc* suppression based on the interaction between the HPV-transforming protein E7 and pRB. Resistance to TGF-β_1-mediated growth inhibition was confirmed for HPV-16-containing tumorigenic keratinocytes by Woodworth et al. [60] and Braun et al. [61]. For HPV-16- [60, 61] or HPV-18- [61] immortalized nontumorigenic keratinocytes, however, both authors describe growth inhibition after treatment with TGF-β_1 correlating with decreased viral transcription. This indicates that the presence of transcriptionally active HPV 16 or 18 DNA alone does not confer resistance to this cytokine. The different responses of nontumorigenic and tumorigenic HPV-positive keratinocytes on TGF-β_1 suggest that this factor may lead to regression of preneoplastic HPV-induced lesions in early stages of infection. Loss of responsiveness to TGF-β accompanied by an unregulated expression of c-*myc* and HPV-transforming genes in the particular keratinocytes might be an important step in the progression of HPV-associated neoplasia.

Cytokines Produced by Immunocompetent Cells

The cytokine network built up by immunocompetent cells infiltrating the particular tissue obviously plays an important role in modulating HPV infection.

It has been reported that treatment with interferons, especially interferon (IFN)-α induces regression or prevents recurrence of HPV-induced lesions [62–67]. In vitro, IFN-α and IFN-γ have been shown to inhibit HPV 18 gene expression in HeLa cells [68]. A more detailed investigation of the effect of IFN on HPV led to contradictory results. Khan et al. [69] found that low concentrations of IFN-α were necessary to inhibit the growth of HPV-16-containing keratinocytes. This effect was interpreted as being due to down-regulation of the expression of the HPV 16 E7 protein. In contrast, Woodworth et al. [70] observed only minimal growth reduction of immortalized cells or cervical carcinoma cell lines by recombinant IFN-α and no changes in HPV gene expression. Similar results were obtained for IFN-γ by Kyo et al. [55]. Different mechanisms have to be considered to explain the effectiveness of IFN therapy on HPV-induced lesions including a direct influence on HPV gene expression and a general enhancement of the immune response. Treatment with IFN-γ stimulates expression of major histocompatibility complex (MHC) class I and II molecules in human keratinocytes [71] and, therefore, can result in an enhanced cellular immune response against HPV infection.

Another important cytokine in host control of HPV infection is leukoregulin (LR). Beside antitumor activity, LR stimulates MHC class I gene transcription [70] and increases the sensitivity of HPV-16-immortalized cervical epithelial cells to killing by natural killer (NK) and lymphokine-activated killer (LAK) cells [72]. However, the increase of sensitivity to cytotoxic effects of NK cells is lost when HPV-16-immortalized cells are analyzed at late passage [72]. NK and LAK cells together might play a significant role in the immune response against HPV-infected cells. NK cells may function as immunomodulators releasing substances such as IL-2 [73], IFN-γ [74] and LR [75] which activate and amplify LAK effector cells [72]. A localized immunodepletion and partial loss of responsiveness to these regulatory lymphokines might contribute to malignant conversion of HPV-induced lesions which occurs late after viral infection.

Immune Status of HPV Induced Lesions

Several observations suggest that host immune surveillance and especially the cell-mediated immune response may limit growth and promote regression of HPV-induced lesions. (1) There is a high prevalence of HPV-induced warts and malignant tumors in patients with suppressed cellular immunity [for a review see ref. 76]. (2) The regression of anogenital and skin warts in humans was found to be associated with a pronounced local infiltration of mononuclear cells including cytotoxic T lymphocytes, NK cells and macrophages invading the epidermis and destroying the neoplastic tissue [77–80]. (3) An immune response can be induced against bovine papillomavirus or cottontail rabbit papillomavirus in cattle or rabbits, respectively, by natural infection or vaccination [81–84], and against HPV in experimental approaches [85–90].

Antibodies recognizing early or capsid proteins of papillomaviruses can be demonstrated in animal and human serum [82, 91–94], are elicited in animals through vaccination with recombinant viral proteins [83, 95] and were shown to be neutralizing in different systems [83, 95–97], but they do not seem to play an important role once the viruses have entered the cell.

T Lymphocytes

HPV-induced benign, premalignant and malignant anogenital lesions like CIN and invasive cervical carcinoma are infiltrated with T lymphocytes, the majority of which are CD8 positive [98–101]. Although little is known about the functional properties of these T cells and their importance in the outcome of HPV infection, there is some evidence that they are poorly activated. (1) Immunohistochemical investigation of cervical tumor tissues demonstrated only low numbers of infiltrating T cells expressing IL-2 receptor [102]. (2) Tumor-invading T lymphocytes are characterized by a lack of granzyme expression [F.V. Cromme, pers. commun.]. (3) Ghosh and Moore [102] performed cytotoxicity tests with tumor-infiltrating T lymphocytes isolated from biopsies of patients with cervical cancer using the cervical cancer cell line CaSki and in some cases autologous tumor cells as cytotoxic T lymphocyte (CTL) targets. MHC-restricted killing of autologous tumor cells was observed only after generation of T cell clones. This strongly supports the assumption of an inadequate tumor-specific activation of CTLs.

The experiments of Ghosh and Moore [102] did not clarify whether the T lymphocytes reactive against the autologous tumor recognized an antigen of viral or cellular origin. HPV-specific T cell proliferation could only be shown after in vitro stimulation of peripheral blood lymphocytes from asymptomatic individuals or mice [103–106]. Alternatively, immunization of mice with different HPV antigens elicited specific cytotoxic and helper T cell responses (87, 90, 107, 108) or delayed-type hypersensitivity [89] in vivo.

Macrophages and Langerhans' Cells

Numerous CD1-positive Langerhans' cells with typical dendritic prolongations are present in the lower two-thirds of the stratified epithelium of normal cervical mucosa [98, 109]. The dendrites are able to form a network throughout the epidermis and extend into the stroma. A reduced density of epithelial Langerhans' cells was observed in genital warts and low-grade CIN lesions [99, 100, 110]. In high-grade CIN lesions, the number of Langerhans' cells is further decreased. They are no longer restricted to the basal and intermediate layers of the epithelium but are distributed throughout the entire epithelial thickness and many of them show an abnormal morphology exhibiting shorter and less numerous dendrites. It is possible that HPV-infected epithelium provides unfavorable homing conditions of Langerhans' cells depending on the grade of dedifferentiation associated with the infection [111]. The depletion and morphological alteration of Langerhans' cells may result in ineffective local antigen presentation and, therefore, cause local immunodeficiency, favoring persistent HPV infection of the cervix. The interactions between Langerhans' cells, HPV-infected keratinocytes and effector T cells, however, are still not clarified.

NIH-3T3 cells transfected with the complete early region of the HPV 16 genome or the HPV 16 E7 oncogene are susceptible to killing by activated macrophages [112, 113]. In contrast, nontransfected control cells and cells expressing the HPV 16 E6 oncogene are not lysed [113]. Direct contact between the macrophages and the target cells appears to be necessary for destruction since supernatants derived from activated macrophages are not able to mediate killing of HPV-16-transformed cells [112]. These data imply that macrophages play an important role in host control of HPV infections. Furthermore, the anti-HPV activity of macrophages could possibly explain the effectiveness of IFN-γ as a therapeutic agent against HPV-induced lesions since this cytokine is the most potent macrophage-activating factor.

Expression of MHC and Cell Adhesion Molecules in HPV-Associated Lesions

MHC class I and II molecules as well as intercellular adhesion molecules are cell surface proteins involved in antigen presentation to T lymphocytes [114, 115]. Alterations in expression of these molecules would critically influence the elimination of virus-infected cells. Therefore, different laboratories have examined MHC class I and II expression in biopsies of HPV-induced lesions. Decreased expression of MHC class I molecules or β_2-microglobulin on epithelial keratinocytes could be observed in premalignant cervical lesions [99, 116], in cutaneous warts and to a lesser extent in condylomas and laryngeal papillomas [117]. A high percentage of cervical cancer biopsies is characterized by heterogeneous or complete loss of MHC class I expression [116, 118]. MHC class I down-regulation seems to occur at the posttranscriptional level since mRNAs of the MHC class I genes are present in the cancer cells [116]. The proportion of specimens with altered MHC class I expression increases using allele-specific anti-MHC class I antibodies [118, 119]. Since cytotoxic T lymphocytes are unable to recognize virus-infected cells that do not express MHC class I molecules, down-regulation of MHC class I cell surface expression can be regarded as an escape mechanism from cellular immune surveillance.

MHC class II surface expression is normally restricted to immunocompetent and specialized antigen-presenting cells. In a significant proportion of HPV-associated premalignant and malignant lesions [103, 116, 117, 120], however, MHC class II antigens are focally or uniformly expressed by epithelial cells. It is not yet clear whether this positively or negatively influences the cellular immune response, although MHC-class-II-expressing keratinocytes are principally able to interact with T helper cells and to participate in immunological reactions [121].

The alterations in MHC expression were independent of the presence of HPV DNA [98, 101, 116–118] or the detection of HPV-specific transcripts [116, 120] in the analyzed specimens. In contrast, absence of the transporter protein TAP-1, which is essential for the translocation of immunogenic peptides into the endoplasmic reticulum and is associated with antigen presentation, strongly correlates with a loss of MHC expression in cervical carcinomas [122]. Interestingly, MHC class I down-regulation was found to be increased in cervical cancer metastases when compared with the primary tumor [F.V. Cromme, pers. commun.] suggesting that MHC-negative neoplastic cells have a selective advantage during metastasis.

Coleman et al. [123] reported induction of the intercellular adhesion molecule ICAM I in high-grade CIN lesions. Some of the established cervical cancer cell lines express ICAM I constitutively. Using these cell lines in an in vitro binding assay, ICAM I was shown to enable adhesion of keratinocytes to polyclonally activated T lymphocytes [123]. In many cervical carcinomas, coexpression of ICAM I with MHC class II molecules could be demonstrated [101] on epithelial keratinocytes, rendering them as possible antigen-presenting cells.

The association between the incidence of cervical cancer and particular MHC alleles (HLA-DQ3 and less strongly HLA-DR6) observed by Wank and Thomssen [124] and with limitations by Helland et al. [125] suggests that the effectiveness of immune surveillance of HPV infection is genetically restricted. Interpretation of these studies, however, is inhibited by the low numbers of cases investigated. Moreover, the results were not confirmed by Glew et al. [126].

NK Cells

NK cells have been shown to be present in HPV-infected cervical epithelia in limited amounts [127]. In patients with HPV-16-associated anogenital lesions, NK cell activity against HPV-16-bearing SK-v keratinocytes was significantly reduced, while the number of circulating NK cells as well as their activity towards K562 cells (which are the standard target for NK cells) remained normal [128, 129]. The reduction of natural cytotoxicity was demonstrated to be due to a decreased capacity of the NK cells to specifically bind SK-v cells. It could not be observed in patients with spontaneously regressed or successfully removed lesions and increased with malignant progression. SK-v cells are resistant to the cytotoxic effects of cytokines produced by NK cells like TNF-α or TGF-β [54], suggesting that the reduced SK-v killing is not due to a decreased release of such factors. On the other hand, a soluble factor excreted by the tumor cells seems to be involved in abrogated NK/SK-v cytotoxicity, since supernatants from SK-v cells inhibit the SK-v lysis by NK cells from patients without lesions [128].

In conclusion, information about extracellular control mechanisms of HPV infection is still fragmentary. The development of therapeutic approaches will depend on a more detailed investigation of anti-HPV immune reactions.

References

1 zur Hausen H: Human papilloma viruses and their possible role in squamous cell carcinoma. Curr Top Microbiol Immunol 1976;78:1–30.

2 zur Hausen H: Human papillomaviruses in the pathogenesis of anogenital cancer. Virology 1991;184:9–13.

3 Dürst M, Petrussevska RT, Boukamp P, Fusenig NE, Gissmann L: Molecular and cytogenetic analysis of immortalized human primary keratinocytes obtained after transfection with human papillomavirus 16 DNA. Oncogene 1987;1:251–256.

4 Pirisi L, Yasumoto S, Feller M, Doninger J, DiPaolo JA: Transformation of human fibroblasts and keratinocytes with human papillomavirus type 16 DNA. J Virol 1987;61:1061–1066.

5 Münger K, Phelps WC, Bubb V, Howley PM, Schlegel R: The E6 and E7 genes of the human papillomavirus type 16 are necessary and sufficient for transformation of primary human keratinocytes. J Virol 1989;63:4417–4423.

6 Kaur P, McDougall JK: HPV-18 immortalization of human keratinocytes. Virology 1989; 173:302–310.

7 Hurlin PJ, Kaur P, Smith PP, Perez-Reyes N, Blanton RA, McDougall JK: Progression of human papillomavirus type 18-immortalized human keratinocytes to a malignant phenotype. Proc Natl Acad Sci USA 1991;88:570–574.

8 DiPaolo JA, Woodworth CD, Popescu NC, Notario V, Doninger J: Induction of human cervical cell carcinoma by sequential transfection with human papillomavirus 16 DNA and viral Harvey *ras*. Oncogene 1989;4:395–399.

9 Dürst M, Gallahan D, Jay G, Rhim JS: Glucocorticoid-enhanced neoplastic transformation of human keratinocytes by human papillomavirus type 16 and an activated *ras* oncogene. Virology 1989;173:767–771.

10 Farber E: The multistep nature of cancer development. Cancer Res 1984;44:4217–4223.

11 de Villiers EM, Wagner D, Schneider A, Wesch H, Munz F, Miklaw H, zur Hausen H: Human papillomavirus DNA in women without and with cytological abnormalities: Results of a five-year follow-up study. Gynecol Oncol 1992; 44:33–39.

12 Rappolee DA, Werb Z: Macrophage-derived growth factors. Curr Top Microbiol Immunol 1992;181:87–140.

13 zur Hausen H: Intracellular surveillance of persisting viral infections: Human genital cancer results from deficient cellular control of papillomavirus gene expression. Lancet 1986;ii: 489–491.

14 zur Hausen H: Dichotomous intracellular control of papovavirus infections disrupted in cancer of the cervix. Lancet 1994;343:955–957.

15 Werness BA, Levine A, Howley PM: Association of human papillomavirus types 16 and 18 E6 proteins with p53. Science 1990;248:76–79.

16 Dyson N, Howley PM, Münger K, Harlow E: The human papillomavirus-16 E7 oncoprotein is able to bind to the retinoblastoma gene product. Science 1989;243:934–937.

17 Gage JG, Meyers C, Wettstein FO: The E7 proteins of the nononcogenic human papillomavirus type 6b (HPV-6b) and of the oncogenic HPV-16 differ in retinoblastoma protein binding and other properties. J Virol 1990;64:723–730.

18 Scheffner M, Werness BA, Huibregtse JM, Levine AJ, Howley PM: The E6 oncoprotein encoded by human papillomavirus types 16 and 18 promotes the degradation of p53. Cell 1990; 63:1129–1136.

19 Schwarz JK, Devoto SH, Smith EJ, Chellappan SP, Jakoi L, Nevins JR: Interaction of the p107 and Rb proteins with E2F during the cell proliferation response. EMBO J 1993;12:1013–1020.

20 Chellappan S, Kraus VB, Kroger K, Münger K, Howley PM, Nevins JR: Adenovirus E1A, simian virus 40 tumor antigen, and human papillomavirus E7 protein share the capacity to disrupt the interaction between transcription factor E2F and the retinoblastoma gene product. Proc Natl Acad Sci USA 1992;89:4549–4553.

21 Lam EWF, Morris JDH, Davies R, Crook T, Watson RJ, Vousden KH: HPV 16 E7 oncoprotein deregulates B-*myb* expression: Correlation with targeting of p107/E2F complexes. EMBO J 1994;13:871–878.

22 Bosch FX, Schwarz E, Boukamp P, Fusenig NE, Bartsch D, zur Hausen H: Suppression in vivo of human papillomavirus type 18 E6-E7 gene expression in nontumorigenic HeLa-fibroblast hybrid cells. J Virol 1990;64:4743–4754.

23 Dürst M, Bosch FX, Glitz D, Schneider A, zur Hausen H: Inverse relationship between HPV 16 early gene expression and cell differentiation in nude mouse epithelial cysts and tumors induced by HPV positive human cell lines. J Virol 1991;65:796–804.

24 Drews RE, Chan VTW, Schnipper LE: Oncogenes result in genomic alterations that activate a transcriptionally silent, dominantly selectable reporter gene (neo). Mol Cell Biol 1992;12:198–206.

25 Stewart N, Bacchetti S: Expression of SV 40 large T antigen, but not small t antigen, is required for the induction of chromosomal aberrations in transformed human cells. Virology 1991;180:49–57.

26 Caporossi D, Bacchetti S: Definition of adenovirus type 5 functions involved in the induction of chromosomal aberrations in human cells. J Gen Virol 1990;71:801–808.

27 Hashida T, Yasumoto S: Induction of chromosome abnormalities in mouse and human epidermal keratinocytes by the human papillomavirus type 16 E7 oncogene. J Gen Virol 1991;72:1569–1577.

28 Standbridge EJ, Der CJ, Doersen CJ, Nishimi RY, Peehl DM, Weissman BE, Wilkinson JE: Human cell hybrids: Analysis of transformation and tumorigenicity. Science 1982;215:252–259.

29 Koi M, Morita H, Yamada H, Satoh H, Barrett JK, Oshimura M: Normal human chromosome 11 suppresses tumorigenicity of human cervical tumor cell line SiHa. Mol Carcinog 1989;2:12–21.

30 Saxon PJ, Srivatsan ES, Stanbridge EJ: Introduction of human chromosome 11 via microcell transfer controls tumorigenic expression of HeLa cells. EMBO J 1986;5:3461–3466.

31 Srivatsan ES, Benedict WF, Stanbridge EJ: Implication of chromosome 11 in the suppression of neoplastic expression in human cell hybrids. Cancer Res 1986;46:6174–6179.

32 Jones PA: Altering gene expression with 5-azacytidine. Cell 1985;40:485–486.

33 Rösl F, Dürst M, zur Hausen H: Selective suppression of human papillomavirus transcription in non-tumorigenic cells by 5-azacytidine. EMBO J 1988;7:1321–1328.

34 Crook CP, Morgenstern JP, Crawford L, Banks L: Continued expression of HPV 16 E7 protein is required for maintenance of the transformed phenotype of cells cotransformed by HPV 16 plus EJ-ras. EMBO J 1989;8:513–519.

35 von Knebel Doeberitz M, Rittmüller C, zur Hausen H, Dürst M: Inhibition of tumorigenicity of cervical cancer cells in nude mice by HPV E6-E7 anti-sense RNA. Int J Cancer 1992;51:831–834.

36 Antequera F, Boyes J, Bird AP: High levels of de novo methylation and altered chromatin structure at CpG islands in cell lines. Cell 1990;72:503–514.

37 Rösl F, Achtstätter T, Bauknecht T, Futterman G, Hutter KJ, zur Hausen H: Extinction of the HPV 18 upstream regulatory region in cervical carcinoma cells after fusion with non-tumorigenic human keratinocytes under non-selective conditions. EMBO J 1991;10:1337–1345.

38 Smits HL, Raadsheer E, Rood J, Mehendale S, Slazer RM, van der Noorda J, ter Schegget J: Induction of anchorage-independent growth of human embryonic fibroblasts with a deletion in the short arm of chromosome 11 by human papillomavirus type 16. J Virol 1988;62:4538–4543.

39 Smits PHM, Smits HL, Minnaar RP, Hemmings BA, Mayer-Jaekel RE, Schuurman R, van der Noordaa J, ter Schegget J: The 55 kDa regulatory subunit of protein phosphatase 2A plays a role in the activation of the HPV 16 long control region in human cells with a deletion in the short arm of chromosone 11. EMBO J 1992;11:4601–4608.

40 Hoppe-Seyler F, Butz K: Repression of the human papillomavirus type 18 enhancer by the cellular transcription factor Oct-1. J Virol 1991;65:5613–5618.

41 Bauknecht T, Angel P, Royer HD, zur Hausen H: Identification of a negative regulatory domain in the human papillomavirus type 18 promoter: Interaction with the transcriptional repressor YY1. EMBO J 1992;11:4607–4617.

42 May M, Dong XP, Beyer-Finkler E, Stubenrauch F, Fuchs P, Pfister H: The E6/E7 promoter of extrachromosomal HPV 16 DNA in cervical cancer escapes from cellular repression by mutation of target sequences for YY1. EMBO J 1994;13:1460–1466.

43 Bartsch D, Boye B, Baust C, zur Hausen H, Schwarz E: Retinoic acid-mediated repression of human papillomavirus 18 transcription and different ligand regulation of the retinoic acid receptor β gene in non-tumorigenic HeLa hybrid cells. EMBO J 1992;11:2283–2291.

44 Houle B, Rochette-Egly C, Bradley WEC: Tumor-suppressive effect of the retinoic acid receptor b in human epidermoid lung cancer cells. Proc Natl Acad Sci USA 1993;90:985–989.

45 Kyo S, Inoue M, Nishio Y, Nakanishi K, Akira S, Inoue H, Yutsudo M, Tanizawa O, Hakura A: NF-IL6 represses early gene expression of human papillomavirus type 16 through binding to the noncoding region. J Virol 1993;67:1058–1066.

46 Rollins BJ, Stier P, Ernst T, Wong GG: The human homolog of the JE gene encodes a monocyte secretory protein. Mol Cell Biol 1989;9:4687–4695.

47 Rösl F, Lengert M, Albrecht J, Kleine K, Zawatzky R, Schraven B, zur Hausen H: Differential regulation of the JE gene encoding the monocyte chemoattractant protein (MCP-1) in cervical carcinoma cells and derived hybrids. J Virol 1994;68:2142–2150.

48 Barker JNWN, Milra RS, Griffiths CEM, Dixit VM, Nickoloff BJ: Keratinocytes as initiators of inflammation. Lancet 1991;337:211–214.

49 Luger TA, Schwarz T: Evidence for an epidermal cytokine network. J Invest Dermatol 1990;95:100S–104S.

50 Ristow HJ: A major factor contributing to epidermal proliferation in inflammatory skin diseases appears to be interleukin 1 or a related protein. PNAS 1987;84:1940–1944.

51 Kirnbauer R, Köck A, Schwarz T, Urbanski A, Krutmann J, Borth W, Damm D, Shipley G, Ansel JC, Luger TA: IFN-β_2, B cell differentiation factor 2, or hybridoma growth factor (IL-6) is expressed and released by human epidermal cells and epidermoid carcinoma cell lines. J Immunol 1989;142:1922–1928.

52 Woodworth CD, Simpson S: Comparative lymphokine secretion by cultured normal human cervical keratinocytes, papillomavirus-immortalized, and carcinoma cell lines. Am J Pathol 1993;142:1544–1555.

53 Malejczyk J, Malejczyk M, Urbanski A, Köck A, Jablonska S, Orth G, Luger TA: Constitutive release of IL6 by human papillomavirus type 16 (HPV16)-harboring keratinocytes: A mechanism augmenting the NK-cell-mediated lysis of HPV-bearing neoplastic cells. Cell Immunol 1991;136:155–164.

54 Malejczyk J, Malejczyk M, Köck A, Urbanski A, Majewski S, Hunzelmann N, Jablonska S, Orth G, Luger TA: Autocrine growth limitation of human papillomavirus type 16-harboring keratinocytes by constitutively released tumor necrosis factor-α. J Immunol 1992;149:2702–2708.

55 Kyo S, Inoue M, Hayasaka N, Inoue T, Yutsudo M, Tanizawa O, Hakura A: Regulation of early gene expression of human papillomavirus type 16 by inflammatory cytokines. Virology 1994;200:130–139.

56 Malejczyk J, Malejczyk M, Majewski S, Breitburd F, Luger TA, Jablonska S, Orth G: Increased tumorgenicity of human keratinocytes harboring human papillomavirus type 16 is associated with resistance to endogenous tumor necrosis factor-α-mediated growth limitation. Int J Cancer 1994;56:593–598.

57 Jablonska S, Majewski S, Malejczyk J: Die Immunologie von HPV-Infektionen und der Mechanismus einer latenten Infektion. Hautarzt 1992;43:305–311.

58 Majewski S, Hunzelmann N, Nischt R, Eckes B, Rudnicka L, Orth G, Krieg T, Jablonska S: TGFβ-1 and TNFα expression in the epidermis of patients with epidermodysplasia verruciformis. J Invest Dermatol 1991;97:862–867.

59 Pietenpol JA, Stein RW, Moran E, Yaciuk P, Schlegel R, Lyons RM, Pittelkow MR, Münger K, Howley PM, Moses HL: TGF-β_1 inhibition of c-*myc* transcription and growth in keratinocytes is abrogated by viral transforming proteins with pRB binding domains. Cell 1990;61: 777–785.
60 Woodworth CD, Notario V, DiPaolo JA: Transforming growth factors beta 1 and 2 transcriptionally regulate human papillomavirus (HPV) type 16 early gene expression in HPV-immortalized human genital epithelial cells. J Virol 1990;64:4767–4775.
61 Braun L, Dürst M, Mikumo R, Crowley A, Robinson M: Regulation of growth in human papillomavirus-transformed keratinocytes by transforming growth factor-β: Implications for the control of papillomavirus infection. Mol Carcinog 1992;6:100–111.
62 Strander H, Cantell K: Studies on antiviral and antitumor effect of human leukocyte interferon in vitro and in vivo, in the production and use of interferon for the treatment and prevention of human virus infection. In Vitro Monogr 1974;3:49–56.
63 Haglund S, Lundquist PG, Cantell K, Strander H: Interferon therapy in juvenile laryngeal papillomatosis. Arch Otolaryngol 1982;107:327–332.
64 Eron LJ, Judson F, Tucker S, Prawer S, Mills J, Murphy K, Hickey M, Rogers M, Flannigan S, Hien S, Katz HI, Goldman S, Gottlieb A, Adams K, Burton P, Tanner D, Taylor E, Peets E: Interferon therapy for condylomata acuminata. N Engl J Med 1986;15:1059–1064.
65 Weck PK, Whisnant JK: Therapeutic approaches to the treatment of human papillomavirus diseases. Cancer Cells 1987;5:393–402.
66 Reichman RC, Strike DG: Pathogenesis and treatment of human genital papillomavirus infections: A review. Antiviral Res 1989;1:109–118.
67 Auborn KJ, Steinberg BM: Therapy of papillomavirus-induced lesions; in Pfister H (ed): Papillomavirus and Human Cancer. Boca Raton CRC, 1990, pp 203–223.
68 Nawa A, Nishiyama Y, Yamamoto N, Maeno K, Goto S, Tomoda Y: Selective suppression of human papillomavirus 18 mRNA level in HeLa cells by interferon. Biochem Biophys Res Commun 1990;170:793–799.
69 Khan MA, Tolleson WH, Gangemi JD, Pirisi L: Inhibition of growth, transformation, and expression of human papillomavirus type 16E7 in human keratinocytes by α interferons. J Virol 1993;67:3396–3403.
70 Woodworth CD, Lichti U, Simpson S, Evans CH, DiPaolo J: Leukoregulin and γ-interferon inhibit human papillomavirus type 16 gene transcription in human papillomavirus immortalized human cervical cells. Cancer Res 1992; 52:456–463.
71 Nickoloff BJ, Basham TY, Merigan TC, Morhenn VB: Antiproliferative effects of recombinant γ interferons on cultured human keratinocytes. Lab Invest 1984;51:697–701.
72 Furbert-Harris PM, Evans CH, Woodworth CD, DiPaolo JA: Loss of leukoregulin up-regulation of natural killer but not lymphokine-activated killer lymphocytotoxicity in human papillomavirus 16 DNA-immortalized cervical epithelial cells. J Natl Cancer Inst 1989;81: 1080–1085.
73 Kasahara T, Djeu JY, Dougherty SF, Oppenheim JJ: Capacity of human large granular lymphocytes (LGL) to produce multiple lymphokines: Interleukin-2, interferon, and colony stimulating factor. J Immunol 1983;131:2379–2385.
74 Djeu JY, Stocks N, Zoon K, Stanton GJ, Timonen T, Herberman RB: Positive self regulation of cytotoxicity in human natural killer cells by production of interferon upon exposure to influenza and herpes viruses. J Exp Med 1982; 156:1222–1234.
75 Evans CH, Heinbaugh JA, Ransom JH: Flow cytometric evaluation of leukoregulin as an intrinsic molecular mediator of natural killer lymphocyte cytotoxicity. Lymphokine Res 1987;6:277–297.
76 Benton C, Shahidullah H, Hunter JAA: Human papillomavirus in the immunosuppressed. Papillomavirus Rep 1992;3:23–26.
77 Tay SK, Jenkins D, Maddox P, Hogg N, Singer A: Tissue macrophage response in human papillomavirus infection and cervical intraepithelial neoplasia. Br J Obstet Gynaecol 1987;94: 1094–1097.
78 Tagami H, Ogino A, Takigawa M, Imamura S, Ofuji S: Regression of plane warts following spontaneous inflammation: A histopathological study. Br J Dermatol 1974;90:147–154.
79 Aiba S, Rokugo M, Tagami H: Immunohistologic analysis of the phenomenon of spontaneous regression of numerous flat warts. Cancer 1986;58:1246–1251.
80 Fierlbeck G, Schiebel U, Muller C: Immunohistology of genital warts in different stages of regression after therapy with interferon gamma. Dermatologica 1989;179:191–195.
81 Kreider JW, Bartlett GL: Shope rabbit papilloma-carcinoma complex: A model system of human papillomavirus infections. Clin Dermatol 1985;3:20–26.
82 Jarrett WFH: The natural history of bovine papillomavirus infection. Adv Viral Oncol 1985;5:83–102.
83 Jarrett WFH, Smith KT, O'Neil BW, Gaukroger JM, Chandrachud LM, Grindlay GJ, McGravie GM, Campo MS: Studies on vaccination against papillomaviruses: Prophylactic and therapeutic vaccination with recombinant structural proteins. Virology 1991;184:345–354.
84 Campo MS, Grindlay GJ, O'Neil BW, Chandrachud LM, McGravie GM, Jarrett FH: Prophylactic and therapeutic vaccination against a mucosal papillomavirus. J Gen Virol 1993;74: 945–953.
85 Meneguzzi G, Cerni C, Kieny MP, Lathe R: Immunization against human papillomavirus type 16 tumor cells with recombinant vaccinia viruses expressing E6 and E7. Virology 1991; 181:62–69.
86 Chen L, Thomas EK, Hu SL, Hellström I, Hellström KE: Human papillomavirus type 16 nucleoprotein E7 is a tumor rejection antigen. Proc Natl Acad Sci USA 1991;88:110–114.
87 Chen L, Mizuno MT, Singhal MC, Hu SL, Galloway DA, Hellström I, Hellström KE: Induction of cytotoxic T lymphocytes specific for a syngeneic tumor expression by the E6 oncoprotein of human papillomavirus type 16. J Immunol 1992;148:2617–2621.
88 Chen L, Ashe S, Brady WA, Hellström I, Hellström KE, Ledbetter JA, McGowan P, Linsley PS: Costimulation of antitumor immunity by the B7 counterreceptor for the T lymphocyte molecules CD28 and CTLA-4. Cell 1992;71: 1093–1102.
89 McLean CS, Sterling JS, Mowat J, Nash AA, Stanley MA: Delayed-type hypersensitivity response to the human papillomavirus type 16 E7 protein in a mouse model. J Gen Virol 1993;74:239–245.
90 Feltkamp MCW, Smits HL, Vierboom MPM, Minnaar RP, de Jongh BM, Drijfhout JW, ter Schegget J, Melief CJM, Kast WM: Vaccination with cytotoxic T lymphocyte epitope-containing peptide protects against a tumor induced by human papillomavirus type 16-transformed cells. Eur J Immunol 1993;23:2242–2249.
91 Jochmus-Kudielka I, Schneider A, Braun R, Kimmig R, Koldovsky U, Schneweis KE, Seedorf K, Gissmann L: Antibodies against the human papillomavirus type 16 early proteins in human sera. Correlation of anti-E7 reactivity and cervical cancer. J Natl Cancer Inst 1989; 81:1698–1704.
92 Müller M, Viscidi RP, Sun Y, Guerrero E, Hill PM, Shah F, Bosch FX, Munoz N, Gissmann L, Shah KV: Antibodies to HPV 16 E6 and E7 proteins as markers for HPV-associated invasive cervical cancer. Virology 1992;187:508–514.
93 Galloway DA, Jenison SA: Characterization of the humoral immune-response to genital papillomaviruses. Mol Biol Med 1990;7:59–72.
94 Bonnez W, Kashima HK, Leventhal B, Mounts P, Rose RC, Reichman RC, Shah KV: Antibody response to human papillomavirus (HPV) type 11 in children with juvenile-onset recurrent respiratory papillomatosis (RRP). Virology 1992;188:384–387.
95 Lin YL, Borenstein LA, Selvakumar R, Ahmed R, Wettstein FO: Effective vaccination against papilloma development by immunization with L1 or L2 structural protein of cottontail rabbit papillomavirus. Virology 1992;187:612–619.
96 Christensen ND, Kreider JW, Kan NC, DiAngelo SL: The open reading frame L2 of cottontail rabbit papillomavirus contains antibody-inducing neutralizing epitopes. Virology 1991; 181:572–579.
97 Christensen ND, Kreider JW, Shah KV, Rando RF: Detection of human serum antibodies that neutralize infectious human papillomavirus type 11 virions. J Gen Virol 1992;73:1261–1267.

98 Ferguson A, Moore M, Fox H: Expression of MHC products and leucocyte differentiation antigens in gynaecological neoplasms: An immunohistological analysis of the tumour cells and infiltrating leucocytes. Br J Cancer 1985; 52:551–563.

99 Viac J, Guerin-Reverchon Y, Chardonnet Y, Bremond A: Langerhans cells and epithelial cell modifications in cervical intraepithelial neoplasia: Correlation with human papillomavirus infection. Immunobiology 1990;180: 328–338.

100 Falk S, Falk C, Stutte HJ: Locale Immunreaktionen bei CIN und Zervixkarzinom – Eine histologische und immunhistochemische Studie. Geburtshilfe Frauenheilkd 1991;51:39–44.

101 Glew SS, Duggan-Keen M, Cabrera T, Stern PL: HLA class II antigen expression in human papillomavirus-associated cervical cancer. Cancer Res 1992;52:4009–4016.

102 Ghosh AK, Moore M: Tumour-infiltrating lymphocytes in cervical carcinoma. Eur J Cancer 1992;11:1910–1916.

103 Altmann A, Jochmus-Kudielka I, Frank R, Gausepohl H, Moebius U, Gissmann L, Meuer SC: Definition of immunogenic determinants of the human papillomavirus type 16 nucleoprotein E7. Eur J Cancer 1992;28:326–333.

104 Steele JC, Stancovic T, Gallimore PH: Production and characterization of human proliferative T-cell clones specific for human papillomavirus type 1 E4 protein. J Virol 1993;67: 2799–2806.

105 Strang G, Hickling JK, McIndoe GAJ, Howland K, Wilkinson D, Ikeda H, Rothbard B: Human T cell responses to human papillomavirus type 16 L1 and E6 synthetic peptides: Identification of T cell determinants, HLA-DR restriction and virus type specificity. J Gen Virol 1990;71:423–431.

106 Stauss HJ, Davies H, Sadovnikova E, Chain B, Horowitz N, Sinclair C: Induction of cytotoxic T lymphocytes with peptides in vitro: identification of candidate T-cell epitopes in human papilloma virus. Proc Natl Acad Sci USA 1992;89:7871–7875.

107 Zhou J, McIndoe A, Davies H, Sun XY, Crawford L: The induction of cytotoxic-T-lymphocyte precursor cells by recombinant vaccinia virus expressing human papillomavirus type 16 L1. Virology 1991;181:203–210.

108 Shepherd PS, Tran TTT, Rowe AJ, Cridland JC, Comerford SA, Chapman MG, Rayfield LS: T cell responses to the human papillomavirus type 16 E7 protein in mice of different haplotypes. J Gen Virol 1992;73:1269–1274.

109 Morelli AE, Sananes C, Di Paola G, Paredes A, Fainboim L: Relationship between types of human papillomavirus and Langerhans' cells in cervical condyloma and intraepithelial neoplasia. Am J Clin Pathol 1993;99:200–206.

110 Hughes RG, Norval M, Howie SEM: Expression of major histocompatibility class II antigens by Langerhans' cells in cervical intraepithelial neoplasia. J Clin Pathol 1988;41:253–259.

111 Fink-Puches R, Smolle J: Langerhans cells in epithelial skin tumors: A quantitative immunohistological and morphometric investigation. In Vivo 1993;7:213–216.

112 Denis M, Chadee K, Matlashewski GJ: Macrophage killing of human papillomavirus type 16-transformed cells. Virology 1989;170: 342–345.

113 Banks L, Moreau F, Vousden K, Pim D, Matlashewski G: Expression of the human papillomavirus E7 oncogene during cell transformation is sufficient to induce susceptibility to lysis by activated macrophages. J Immunol 1991;146:2037–2042.

114 Tanaka K, Yoshioka T, Biebnerich C, Jay G: Role of major histocompatibility complex class I antigens in tumor growth and metastasis. Annu Rev Immunol 1988;6:359–380.

115 Springer TA: Adhesion receptors of the immune system. Nature 1990;346:425–434.

116 Cromme FV, Meijer CJLM, Snijders PJF, Uyterlöinde A, Kenemans P, Helmerhorst T, Stern PL, van den Brule AJC, Walboomers JMM: Analysis of MHC class I and II expression in relation to presence of HPV genotypes in premalignant and malignant cervical lesions. Br J Cancer 1993;67:1372–1380.

117 Viac J, Soler C, Chardonnet Y, Euvrard S, Schmitt D: Expression of immune associated surface antigens of keratinocytes in human papillomavirus-derived lesions. Immunobiology 1993;188:392–402.

118 Connor ME, Stern P: Loss of MHC class-I expression in cervical carcinomas. Int J Cancer 1990;46:1029–1034.

119 Duggan-Keen M, Keating PJ, Cromme FV, Walboomers JMM, Stern PL: Alterations in major histocompatibility complex expression in cervical cancer. Possible consequences for immunotherapy. Papillomavirus Rep 1994;5: 3–9.

120 Jochmus I, Dürst M, Reid R, Altmann A, Bijward K, Gissmann L, Jenson AB: Major histocompatibility complex and human papillomavirus type 16 E7 expression in high-grade vulvar lesions. Hum Pathol 1993;24:519–524.

121 Cunningham AL, Noble JR: Role of keratinocytes in human recurrent herpetic lesions: Ability to present herpes simplex virus antigen and act as targets for T lymphocyte cytotoxicity in vitro. J Clin Invest 1989;83:490–496.

122 Cromme FV, Airey J, Heemels MT, Ploegh HL, Keating PJ, Stern PL, Meijer CJLM, Walboomers JMM: Loss of transporter protein, encoded by the TAP-1 gene is highly correlated with loss of HLA expression in cervical carcinomas. J Exp Med 1994;179:335–340.

123 Coleman N, Greenfield IM, Hare J, Kruger-Gray H, Chain BM, Stanley MA: Characterization and functional analysis of the expression of intercellular adhesion molecule-1 in human papillomavirus-related disease of cervical keratinocytes. Am J Pathol 1993;143: 355–367.

124 Wank R, Thomssen C: High risk of squamous cell carcinoma of the cervix for women with HLA-DQw3. Nature 1991;352:723–725.

125 Helland A, Borresen AL, Ronningen KS, Thorsby E: HLA antigens and cervical carcinoma. Nature 1992;356:23.

126 Glew SS; Stern PL, Davidson JA, Dyer PA: HLA antigens and cervical carcinoma. Nature 1992;356:22.

127 Tay SK, Jenkins D, Singer A: Natural killer cells in cervical intraepithelial neoplasia and human papillomavirus infection. Br J Obstet Gynaecol 1987;94:901–906.

128 Malejczyk J, Majewski S, Jablonska S, Rogozinski TT, Orth G: Abrogated NK-cell lysis of human papillomavirus (HPV)-16-bearing keratinocytes in patients with precancerous and cancerous HPV-induced anogenital lesions. Int J Cancer 1989;43:209–214.

129 Malejczyk J, Malejczyk M, Majewski S, Orth G, Jablonska S: NK-cell activity in patients with HPV16-associated anogenital tumors: Defective recognition of HPV16-harboring keratinocytes and restricted unresponsiveness to immunostimulatory cytokines. Int J Cancer 1993;54:917–921.

Intervirology 1994;37:189–200

Janet L. Brandsma

Section of Comparative Medicine and Department of Epidemiology and Public Health, Yale University School of Medicine, New Haven, Conn., USA

Animal Models of Human-Papillomavirus-Associated Oncogenesis

Key Words
Animal model
Papillomavirus
Pathogenesis
Papilloma
Neoplasia
Oncogenesis
Replication
Genetics
Mutational analysis

Introduction

Cervical intraepithelial neoplasia (CIN) is a precancerous lesion caused by infection with human papillomavirus (HPV). CIN spontaneously regresses in some women, but in other persists for years. The longer it persists, the greater the likelihood that it will progress to cervical cancer [1; A. Schneider, this issue]. Treatment of CIN relies on therapeutic modalities ranging from cryotherapy and surgical excision to topical treatments with podophyllin and trichloroacetic acid [2; L. Cowsert, this issue]. No therapy is uniformly effective, and even for patients who do respond, recurrence is often a problem. Destruction of all infected tissue does not appear to be technically feasible, since multifocal disease and latency are common features of infection. Antiviral drugs that target viral replication are unlikely to be effective because the virus does not replicate in the cells that maintain the infection (basal keratinocytes) [T.R. Broker and L.T. Chow, this issue] and, in high-grade CIN, viral genes involved in replication are typically lost through integration, rearrangement and deletion [1; R. Schlegel, this issue].

Need for an Animal Model

To expedite the development of prophylactic and therapeutic strategies, we need to understand how viral genes contribute to papilloma formation and, subsequently, to malignant progression. For this review, the term 'papilloma' describes a variety of benign proliferative epithelial lesions that are characteristic of papillomavirus infections. While great progress has been made recently in understanding the functions of oncogenes and regulatory genes of papillomaviruses in cultured cells [3–7], relatively little is known about the roles of viral genes in inducing benign lesions and mediating neoplastic progression. Such information can best be gained by experiments in animals.

It has long been a goal of papillomavirus research to establish an animal model of HPV-induced disease to better characterize acute infection as well as invasive and metastatic carcinoma. A model would also be used to define and characterize genetic and environmental cofactors that accelerate or impede disease progression. It would facilitate investigation of the molecular mecha-

Janet L. Brandsma, PhD
Section of Comparative Medicine
P.O. Box 208016
New Haven, CT 06520-8016 (USA)

0300–5526/94/0374–0189
$8.00/0

nisms that underlie pathogenic processes, and result in identification of molecular targets for therapeutic intervention. It would be useful to develop prophylactic and therapeutic strategies, and would also provide the means to evaluate experimental therapies prior to clinical trials.

Desirable Features of an Animal Model

Experimental disease should mimic the human disease in essential aspects of etiology, pathology and natural history. Ideally, it would be caused by a papillomavirus, preferably a high-risk human papillomavirus, i.e. one associated with human cancer or, at least, by papillomavirus genes. The use of recombinant viral DNA would have an advantage over virions, because DNA can be manipulated to generate specific mutants that would be exceedingly valuable for characterizing viral gene functions. The induced disease would affect mucosa or skin, the natural targets of HPV infection. Benign proliferative lesions would develop early and, within a reasonable time frame, a significant proportion would progress to squamous cell carcinoma. The host would be a small and affordable laboratory species, or (for some studies) a nonhuman primate.

Sources of Virus and Susceptible Hosts

An animal model of a viral disease typically requires a source of infectious virus and a susceptible host. Both aspects have posed problems for an HPV model. Before suitable cell culture systems were developed, inoculated animals were often used for growing human viruses. For example, rabies virus was routinely grown in suckling-mouse brain [8]. Since HPVs are species specific, however, growing them in laboratory animals is unlikely to succeed. Moreover, a suitable in vitro system for propagating papillomaviruses has not yet been defined. This is because keratinocytes, the natural hosts and probably the only unmodified cells in which papillomaviruses replicate, produce viral progeny only when they are highly differentiated. Culture conditions to achieve that state have been difficult to develop [T.R. Broker and L.T. Chow, this issue]. It has recently been reported that HPV replication could be achieved in raft cultures, using human keratinocytes that already harbored HPV [3, 10; T.R. Broker and L.T. Chow, this issue]. Because the cultured keratinocytes were not infected in vitro, the raft culture approach does not enable isolation of HPVs from clinical specimens. In addition, virion production is low, and no virus stocks have been generated.

Another potential source of genital HPV might be biopsies of naturally infected human genital tissue. Infected human tissue (blood) is used, for example, to isolate the human parvovirus B19 [11]. Extraction of HPV from human genital lesions is, however, unlikely to yield sufficient virus for experimental studies because genital lesions, particularly those infected with high-risk virus types, seem to produce little virus [12]. One exception is low-risk HPV 11, fortuitously isolated from a pool of condylomata acuminata (described below) [13].

A final possibility would be to use viral DNA instead of whole virus. In the early 1960s, it was discovered that naked DNA, isolated from papillomas or carcinomas induced by the cottontail rabbit papillomavirus (CRPV), could induce papillomas on naive rabbits [14–16]. This finding was extended to recombinant CRPV DNA by Nasseri and Wettstein [17], and has been reproduced in other laboratories, using independent clones of CRPV [18, 19]. This work implies that similar results might be obtained with HPV DNAs, if a wild-type clone were inoculated into a susceptible host. An obvious advantage of using recombinant DNA methods is that they permit the generation of viral mutants. The ability to assay biological and biochemical functions of mutant genes has been central to investigation of viral gene functions in vitro, and would also be essential for studies in vivo, i.e. in whole animals.

Apart from the difficulties of obtaining HPV virus is the additional problem of identifying a susceptible host. Despite early reports that extracts of human lesions could induce disease in dogs, monkeys and rabbits [20], none of those models has been reproducible [21], and it appears that HPVs infect only human beings. Species-specific barriers may reside in viral capsid proteins that mediate viral cell entry. They are also likely to involve viral regulatory sequences and proteins that effect transcription, translation and replication by interaction with intracellular proteins, which may be inappropriate or insufficient in the wrong host. Three approaches have been utilized to circumvent the problem of species specificity. In a homologous approach, a nonhuman papillomavirus is studied in its natural nonhuman host. In a heterologous approach, HPV genomes (or genes) are incorporated into the germ line of heterologous species. In a chimeric approach, the host (or the virus) is manipulated to allow biological interaction with selected viruses (or hosts).

Table 1. Cloned isolates of naturally occurring nonhuman mammalian papillomaviruses

Common animal name	Genus and species	Virus name	Site of isolation	Evidence of cancer[1]	Experimental disease[2]	Reference[3]
Rat[4]	*Mastomys natalensis*	MnPV	skin (epidermis)	yes	yes	23
Mouse	*Micromys minutus*	MmPV	skin (epidermis)	yes	no[5]	24
Rabbit	*Sylvilagus floridensis*	CRPV	skin (epidermis)	yes	yes	25–27
Rabbit	*Oryctolagus cuniculus*	CRPV	skin (epidermis)	yes	yes	25–27
Rabbit	*Oryctolagus cuniculus*	ROPV	oral cavity (mucosa)	no	yes	28
Cattle	*Bos taurus*	BPV 1	skin (epidermis and dermis)	no	yes	29, 30
Cattle	*Bos taurus*	BPV 4	alimentary tract (mucosa)	no	yes	31
Horse	*Equus caballus*	EqPV	skin (epidermis)	no	yes	32
Dog	*Canis familiaris*	COPV	oral cavity (mucosa)	yes	yes	33
Monkey	*Macaca mulatta*	RhPV 1	lymph node (metastasis[6])	yes	no	34
Monkey	*Colobus guereza*	CgPV 2	skin (epidermis)	no	no	34
Monkey	*Colobus guereza*	CgPV 1	penis (epidermis)	no	no	35
Deer	*Odocoileus virginianus*	DPV	skin (epidermis and dermis)	no	no	36
Elk[7]	*Alces alces*	EEPV	skin (epidermis and dermis)	no	no	37
Reindeer	*Rangifer tarandus*	RPV	skin (epidermis and dermis)	no	no	38

1 Evidence of naturally occurring malignant disease.
2 Disease has been reproduced experimentally.
3 Reference for molecular cloning of the viral genome.
4 Multimammate rat.
5 With the exception of one wild harvest mouse.
6 Of a penile squamous cell carcinoma.
7 European elk.

Animal Papillomaviruses in Their Natural Hosts

The homologous approach is straightforward. The use of an animal papillomavirus in its natural host removes the problems of virus availability (virus is propagated in the animal) and species specificity (no human elements are involved). Numerous naturally occurring mammalian papillomaviruses have been described, and several have been isolated (table 1) [22]. In many cases, the viral infection has not been associated with cancer or the disease has not been experimentally transmitted. Some species are poorly suited for experimental studies because they are large and expensive, e.g. cattle and horses, and some are, additionally, difficult to obtain and maintain, e.g. wild deer, elk, reindeer. Animal papillomaviruses that have been or might be used in animal models affect rodents, cattle, rabbits, dogs and nonhuman primates.

Rodents

The multimammate rat *Mastomys natalensis,* and the wild European harvest mouse *Micromys minutus* have papillomaviruses, but are not common laboratory species. *M. natalensis* papillomavirus (MnPV) has been experimentally transmitted to young *Mastomys* using tumor extracts. In addition, episomal MnPV DNA has been identified, in an apparently latent state, in normal *Mastomys* tissues [23, 39]. Endogenously affected *Mastomys* underwent spontaneous amplification of MnPV DNA in cells of the epidermis as the animals aged, and amplification was accompanied by development of tumors. Histologically, the tumors were most frequently keratoacanthomas, sometimes papillomas, and occasionally keratinizing squamous cell carcinomas. Activation of MnPV genomes in normal skin could be potentiated by treatment with the tumor promoter 12-O-tetradecanoylphorbol-13-acetate (TPA) [23] or by chronic mechanical irritation [40]. Still, the endogenous nature of MnPV is clearly distinct from that of HPV. Another endogenous papillomavirus that can be activated to cause papillomas and carcinomas by exposure to ultraviolet light has been reported in hairless mice *(Mus musculus),* but the virus has not been isolated [41, 42]. Little is kown about the natural history of *M. minutus* papillomavirus (MmPV) infection, although a variety of lesions, including hyperplasia, epidermal inclusion cysts, squamous papillomas, inverted papillomas, trichoepitheliomas and sebaceous carcinomas have been associated with the virus in wild harvest mice [43]. Further study of MmPV is complicated by the inability to transmit the virus to laboratory mice [44].

Cattle

Warts induced by bovine papillomavirus (BPV) are a significant agricultural problem in the United States, because animals affected with a communicable disease may not cross state lines [45]. BPV types 1, 2 and 5 cause fibropapillomas, lesions with dermal as well as epidermal components [22, 46]. There is no evidence that BPV-induced fibropapillomas ever undergo malignant progression during natural infection. Thus, the pathology and natural history of this disease differ from human cervical disease, which involves only epidermis and may become malignant. Nevertheless, BPV 1 is relevant for human disease because more is known about the molecular biology of BPV 1 than about any other papillomavirus. This occurred because the first papillomavirus transformation assay was developed using BPV 1. BPV 1 gene functions have been extensively characterized in vitro, using mutagenesis analysis [47]. It would be desirable to expand such findings to whole animals, a possibility if a means to generate infectious BPV 1 mutants can be developed. BPV types 3, 4 and 6 cause epithelial papillomas [22]. BPV 4 infects the alimentary tract, where the papillomas can progress to squamous cell carcinoma. However, the oncogenic process in cattle appears not to involve the continued presence of BPV 4 genes and thus is unlike HPV-associated oncogenesis [48].

Rabbits

Richard Shope [49] identified the first DNA tumor virus in 1933. Known as the Shope papillomavirus (or CRPV), the virus induces papillomas in rabbit skin that are histologically similar to cutaneous human papillomas. CRPV can be experimentally transmitted to cottontail *(Sylvilagus)* and domestic *(Oryctolagus)* rabbits. This system was studied extensively in the decades folllowing its discovery, and has provided important insights into the pathogenesis of papillomavirus-associated disease.

The natural history of CRPV-induced disease parallels that of high-risk HPV infection [50, 51]. Lesions develop a few weeks after experimental infection. Virus replicates in the differentiated layers of cottontail rabbit papillomas, but not in basal or parabasal cells [52]. Spontaneous regression occurs due to cell-mediated immune mechanisms, but papillomas that persist can undergo malignant conversion to squamous cell carcinomas [53, 54]. Chemical carcinogens, such as tar, methylcholanthrene or benzanthracene, increase the percentage of malignant conversions and shorten the latency period [55, 58]. Malignant conversion of CRPV-induced papillomas occurs in up to 75% of domestic rabbits, beginning about 6 months after infection and continuing for up to 18 months [58, 59]. More recently, it has been shown that the frequency of malignant progression in rabbits is associated with major histocompatibility complex alleles [60], as has also been suggested for cervical cancer [61]. The CRPV-rabbit system has been employed for evaluation of experimental therapies. Kreider et al. [62] showed that 9-(2-phosphonylmethoxy) ethyl-guanine (PMEG) was useful for treatment of latent CRPV infection in rabbits, as it was for latent HPV infection in humans, a result that again emphasizes the similarities between the two systems. Ostrow et al. [63] have shown that the nucleoside analog ribavirin can impede the growth of early warts caused by CRPV. Collectively, these features underscore the relevance of the CRPV system for HPV-associated human cancer, and make it attractive for experimental studies.

A special strength of the CRPV-rabbit system is that papillomas and carcinomas [17–19, 64] can be induced with molecular clones of the viral genome, making it currently the only system in which mutagenesis studies can be performed in vivo. The major obstacle to using CRPV DNA for tumor induction were the tedious inoculation methods that yielded few papillomas per animal. We introduced a jet injector to deliver CRPV DNA and showed that each animal could be inoculated at ≥100 sites in 10 min, making the method far superior to previous methods [18]. The complete life cycle of CRPV, including production of viral progeny, can be induced with a cloned CRPV genome in the natural host [64]. CRPV DNA-induced papillomas can also progress to squamous cell carcinoma [J.L. Brandsma et al. unpubl. data]. Thus, this unique system can be used to elucidate viral gene functions that control papilloma formation, virus production and malignant conversion (fig. 1).

CRPV genomes harboring site-specific mutations have been tested in domestic rabbits, and results revealed that the E1, E2, E6, E7 and L1 genes are each essential for papilloma formation (table 2) [18, 65–68]. Mutant CRPV genomes containing a defective E1, E2 or E6 frameshift mutation could be rescued by replacing each mutation with its wild-type counterpart. The marker rescue mutants, when tested in rabbits, displayed wild-type activity, confirming that the loss of papilloma-inducing ability was due to the engineered mutation [64a]. CRPV genomes harboring mutations in the putative E5 gene, in contrast, did not abolish papilloma formation, although the frequency was generally lower, suggesting that the E5 gene may facilitate rather than be required for papilloma formation [67]. It is possible that in cottontail rabbits the CRPV E5 gene is required for virus production.

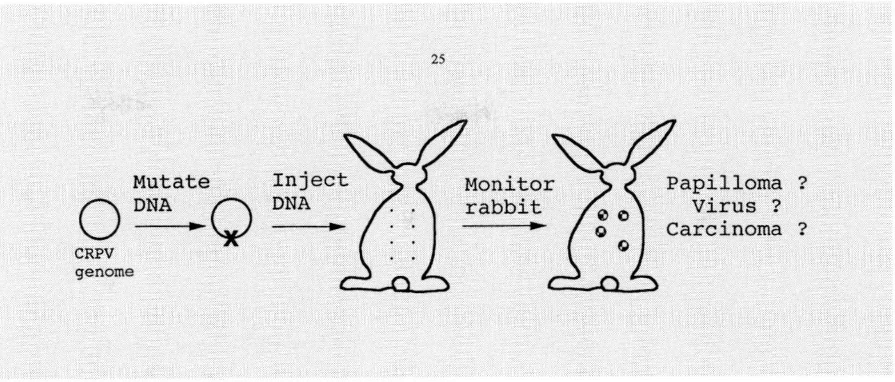

Fig. 1. Papillomavirus mutagenesis analysis in vivo.

Table 2. CRPV gene requirements for papilloma formation in domestic rabbits

Gene[1]	Nature of mutation[2]	Papilloma formation	Reference
E1	frameshift (3)	0	64a
	rescue (1)	+	64a
E2	frameshift (2)	0	64a
	rescue (1)	+	64a
E5	codon, deletion, frameshift (3)	±	39, 65
E6	deletion, codon, frameshift (3)	0	64a, 65
	rescue (1)	+	64a
E7	frameshift (2)	0	18, 65
L1, L2	deletion (1)	0	66
L2	deletion (1)	+	66

1 Genes were cloned in a CRPV genome in plasmid pLAII and inoculated as recombinant plasmids.
2 Codon = altered start codon; rescue = marker rescue. In parentheses, the number of different mutations that were tested.

The CRPV E7 gene has been studied in greater detail. The encoded protein binds the retinoblastoma suppressor protein (pRB) and has the associated properties of disrupting pRB/E2F/DNA complexes and transactivating the adenovirus E2 promoter, that have been described for the HPV 16 E7 protein [68, 69]. The in-frame mutant CRPV-E7p9Leu contained an additional leucine residue at amino acid 9 and encoded a full-length protein with pRB-binding and associated properties characteristic of the wild-type protein. Intradermal inoculation of rabbits with CRPV-E7p9Leu, however, did not induce lesions, suggesting that this region encodes a critical function [18]. The region has homology to the 300 kD protein-binding domain of adenovirus E1A protein, and it has been speculated that E1A-300 kD interactions might down-regulate transcription of cellular genes important for cell growth. CRPV E7 protein could function in a similar fashion [18].

Two additional CRPV mutant genes, E7($C^{24}\rightarrow A^{24}$) and E7($L^{22}\rightarrow A^{22}$, $C^{24}\rightarrow A^{24}$), have been constructed at the site that encodes pRB-binding activity. E7($C^{24}\rightarrow A^{24}$) and E7($L^{22}\rightarrow A^{22}$, $C^{24}\rightarrow A^{24}$) proteins had a considerably or profoundly reduced ability, respectively, to bind pRB, compared to wild-type CRPV or HPV 16 E7 [68, 69]. CRPV genomes containing the E7($L^{22}\rightarrow A^{22}$, $C^{24}\rightarrow A^{24}$) or the E7($C^{24}\rightarrow A^{24}$) gene induced papillomas at 3% of inoculated sites, a frequency that was significantly lower than wild-type CRPV (19%) (table 3) [68]. These results showed that papilloma induction does not absolutely require direct interaction between CRPV E7 protein and

Table 3. Papilloma formation by CRPV genomes containing in-frame E7 mutations

Mutant E7 gene[1]	Location of mutation	Relative pRB binding[2]	Papilloma formation[3]		Reference
			absolute %	relative %	
E7p9Leu	codon 9	60.0	0	0	18, 68
E7p9Leu Rsc	rescue of E7p9Leu	ND	45	128	64a
E7(C24→A24)	codon 24	1.2	3	16	68
E7(C22→A22, C24→A24)	codons 22+24	≤0.1	3	16	68

ND = Not done.

1 E7 mutant genes were cloned in a CRPV genome in plasmid pLAII and inoculated as recombinant plasmids.

2 Percentage of pRB-binding activity relative to CRPV E7. Determined in a competitive binding assay for human pRB in the presence of HPV16 E7 protein.

3 Percentages represent the average of results for each mutant. Values are the average of all results. Absolute = Percentage of sites inoculated that formed papillomas; relative = relative to wild-type CRPV DNA.

pRB, and suggest that properties of E7 other than binding pRB are required. It remains to be elucidated whether E7-pRB binding is involved in malignant progression.

Another papillomavirus of rabbits is the rabbit oral papillomavirus (ROPV) [70]. ROPV causes small papillomas of the oral cavity in domestic rabbits. Large amounts of ROPV are produced in oral lesions, but the virus cannot be transmitted to the conjunctiva or lower genital tract. Although oral papillomas are common in some rabbit colonies and persist for long periods, there is no evidence of malignant progression to squamous cell carcinoma [22].

Dogs

Canine oral papillomavirus (COPV) causes papillomas on the buccal mucosa, dorsal tongue and gingiva, but the lesions do not undergo malignant progression [22]. If, however, dogs are inoculated intramuscularly with COPV, they develop skin lesions at the site of inoculation, including hyperplasia, cysts, squamous papilloma, basal cell epithelioma, and squamous cell carcinoma [22, 34, 71, 72]. If it can be established that COPV was the cause of the carcinomas, this would be a potential model for investigating papillomavirus-mediated oncogenesis, although the incidence of carcinoma is low and the incubation period is 2–3 years.

Nonhuman Primates

Papillomaviruses have been cloned from a cutaneous and a penile lesion on colobus monkeys [35, 36, 73] and from a lymph node metastasis on a rhesus monkey [36]. A fourth papillomavirus has been described in an oral lesion on a pygmy chimpanzee, but the virus has not been isolated [74]. Only the rhesus papillomavirus (RhPV 1) has been associated with cancer, and the primary carcinoma that metastasized to the lymph node from which RhPV 1 was isolated was a penile carcinoma. Since penile carcinomas in men are strongly associated with high-risk genital HPVs, RhPV-1-induced genital tract disease is potentially relevant and merits further consideration. A retrospective study of sexually interactive rhesus monkeys revealed the presence of multiple RhPV-1-positive neoplastic genital lesions of varying severity, including squamous cell carcinoma of the cervix [75]. Thus, the pathogenesis and natural history of RhPV 1 infection in rhesus monkeys may provide an excellent model for HPV-associated cervical oncogenesis. The RhPV 1 genome has been sequenced and is most closely related to high-risk HPV 16 [76]. RhPV 1 E5, E6 and E7 genes, like their HPV 16 counterparts, transform cells in vitro [77, 78]. Thus far, however, experimental disease has not been achieved with RhPV 1 because of the lack of a source of infectious virus. Theoretically, it should be possible to induce disease by molecular inoculation of rhesus monkeys with cloned RhPV 1 genomes, as was done in the CRPV-rabbit system. A wild-type genomic clone is likely to be required, but it is not known whether the prototype RhPV 1 genome, obtained from a metastasis, contains mutation(s) that would preclude the ability to induce disease. Prototype HPV 16, for example, that was cloned from a cervical carcinoma, contains crippling mutations in the E1 [79, 80] and L1 [81] genes.

Table 4. Papillomavirus transgenic animals

Transgene	Promoter	Site of expression	Pathology	Latency period	Reference
BPV 1 genome	endogeneous	skin (dermis)	fibromatosis →fibrosarcoma	8–9 months	82–85
CRPV genome	endogeneous	skin (epidermis)	papilloma	1 month	86
CRPV genome + EJ-*ras*	endogeneous	skin (epidermis)	squamous cell carcinoma	'early'	86
HPV 1e early region	keratin 6	skin (epidermis)	abnormal differentiation	7 days	87
HPV 16 E6/E7	MMTV LTR	testes (germ cells)	seminoma	7 months	88
HPV 16 E6/E7	β-actin	brain (neuroepithelium)	anaplastic and well-differentiated carcinoma	2.5 months	89
HPV 16 E6/E7	αA-crystallin	ocular lens (epithelium)	microphthalmia and cataracts →locally invasive malignant tumor	6–12 months	90
HPV 16 E6/E7	αA-crystallin	skin (epithelium)	acanthosis, hyperkeratosis and hair loss →squamous cell carcinoma	3 months	91
HPV 16 E6/E7	keratin 10	skin (epithelium)	hyperplasia →papilloma, cancer	2–12 months	L. Gissmann [pers. commun.]
HPV 16 early region	keratin 14	skin (epithelium)	hyperplasia →dysplasia, papilloma, carcinoma	1–10 months	92

Transgenic Animals

Another strategy to circumvent the species specificity of papillomaviruses is to transfer viral genes to animals as heritable transgenes, thereby bypassing the requirements for infectious virus and inherent susceptibility to infection. In an optimal transgenic model for papillomavirus-associated oncogenesis, expression of the papillomavirus transgenes would cause benign proliferative mucosal or cutaneous lesions that progress to squamous cell carcinoma. Transgenic models can be used to characterize the contributions of viral genes, cellular elements and environmental cofactors to oncogenesis, and to assess their relative importance. Transgenic mice have been engineered to carry the entire BPV 1 genome, the early region of HPV 1e, and the E6 and E7 genes of HPV 16 (table 4) [82–85, 87–91]. Transgenic rabbits that carry CRPV with or without an activated *ras* oncogene have also been engineered (table 4) [86].

The first papillomavirus transgenic mice were established by Hanahan et al. [83]. They carried the BPV 1 genome and developed dermal lesions involving distinct proliferative stages, beginning with a long latency period, followed by development of benign fibromatosis, and finally progression to locally invasive fibrosarcomas. Fibromatosis was accompanied by extrachromosomal replication of viral DNA [83, 84]. The fibrosarcomas, but not the precursor lesions, displayed characteristic karyotypic abnormalities, indicating that additional genetic alterations occur during tumorigenesis [85] and that the model might prove useful for identification of the cellular genes that are involved.

Peng et al. [86] produced two CRPV transgenic rabbits that carried the CRPV genome. Both developed extensive cutaneous papillomas at about 1 month of age. Transcription of CRPV was restricted to skin, indicating that viral gene expression is not limited by cellular access, but by tissue-specific factors. A third transgenic rabbit carried, in addition to CRPV, an activated EJ-ras oncogene [87]. This animal developed extensive skin carcinomas at a very early age, suggesting a critical role for the *ras* oncogene. None of the three CRPV-transgenic rabbits became founders, however, and no lineage was established for further study.

Three laboratories have produced transgenic mice that carry the E6 and E7 oncogenes of HPV 16 [88–91]. Each lineage provides an independent in vivo system to investigate the mechanisms by which E6 and E7 genes induce the observed pathology, and how cellular genes interact with viral genes in tumorigenesis. Mice established by Kondoh et al. [88] expressed E6/E7 transgenes from the mouse mammary tumor virus promoter and developed large bilateral testicular seminomas. E6 and E7 mRNAs were expressed at high levels in all tumors, and also in normal submandibular gland, but not in lung or spleen. Three transgenic lineages were established and, among them, the incidence of tumorigenesis varied from 17 to 100%, suggesting that additional cellular genetic alterations were involved in the malignant progression.

The β-actin promoter was used by Arbeit et al. [89] to express the HPV 16 E6/E7 genes in transgenic mice. Malignant neuroepithelial tumors developed in the brains of these mice, killing the majority of animals within 10 months. Two thirds of the tumors were anaplastic, and one third were well differentiated. E6/E7 mRNA and E7 protein expression were specifically found in both tumor types. The p53 and pRB oncogenes, to which E6 and E7 bind, respectively, appeared to be wild type in the tumor cells. An advantage of this system is that the latency period is relatively short, with abnormal hyperplastic cells appearing as early as 1 month and primitive tumors at 2.5 months.

The αA-crystallin promoter has been used by Lambert et al. [91] to produce transgenic mice carrying HPV 16 E6/E7 genes. Epithelial cells of the ocular lens expressed the transgenes, as expected, and bilateral microphthalmia and cataracts developed in all mice. In one lineage, lens tumors developed from the benign lesions in 40% of the animals. The tumors destroyed internal ocular structures, which was interpreted as evidence of local malignancy. In culture, primary neonatal lens grew as immortalized cells and acquired a tumorigenic phenotype after passage, indicating that cellular genetic changes in addition to the transgenes are probably necessary for tumorigenesis.

Expression of the αA-crystallin-E6/E7 transgene also occurred ectopically [91]. Neonatal expression was observed in the epidermis and dermis of skin, which became dry and scaly and was affected by hair loss, beginning at 3 months. Skin abnormalities were significantly more pronounced in homozygotes as compared to heterozygotes, further indicating an etiologic involvement of the transgenes. Locally invasive tumors subsequently developed in 20% of affected animals, after 6 months. The most frequent were squamous cell carcinomas, although fibrosarcomas, a phenotype previously seen in BPV-1-transgenic mice, also occurred occasionally. The tumors had elevated levels of E6/E7 expression, but wild-type p53 and pRB tumor suppressor genes, as has been reported in HPV-associated cervical cancer [6, 7; R. Schlegel, this issue].

More recently, keratin promoters have been used to express HPV 16 early genes in transgenic mice. Expression of the E6 and E7 genes from a suprabasal cell keratin promoter resulted in epidermal hyperplasia, formation of papillomas, and eventually skin cancer [L. Gissman, pers. commun.]. Expression of the entire early region from a basal cell keratin promoter resulted in hyperplasia, dysplasia, papillomas and carcinomas [92].

Chimeric Animals

Although direct HPV infection of heterologous animal species has not succeeded, a model of HPV infection has been developed by Kreider et al. [94] by implanting human genital skin incubated in vitro with HPV 11 virus into athymic (nude) mice. Lesions that are virtually identical to condylomata acuminata develop in about 3 months [94, 95]. This model has proved useful for propagating infectious HPV 11 virus [13], identifying susceptible human target tissues [96], characterizing the dynamics of viral gene transcription during the course of disease induction [96, 97], identifying monoclonal and human neutralizing antibodies to HPV 11 [98, 99], and providing infected tissue for in vitro experiments [9]. It does, however, have some drawbacks. The implants are placed in the renal capsule, so they cannot be monitored. Secondly, only one strain of HPV, isolated from a pool of human condylomata acuminata, induces lesions. Since HPV 11 is not a high-risk type, neoplastic progression does not occur, and the system has not proved useful for study of HPV-associated oncogenesis. Other genital types of HPV appear to be extremely difficult to obtain from human tissues in titers sufficiently high to induce lesions and, in fact, none have been isolated. Finally, conditions have not been found to induce lesions using HPV DNAs, and no viral mutants have been generated or tested.

A nude mouse model developed by Sterling et al. [100] permits induction of HPV-16-positive lesions, but it does not employ infection. This model uses a human cell line (W12) that harbors full-length episomal HPV 16 genomes. Low-passage W12 cells were grafted onto the backs of mice in silicone chambers, where the cells grew and differentiated into a human skin equivalent with some characteristics of a low-grade CIN, including the presence of papillomavirus particles, as visualized by electron microscopy. The amount of virus, however, was small and infectivity was not demonstrated. Although the W12 cell line was established from a low-grade human cervical lesion, the vast majority of cell lines derived from HPV-positive lesions contain only integrated viral DNA, making it unlikely that the model will have widespread applicability.

A third approach is being explored in our laboratory: the use of HPV DNA to induce lesions. Because molecu-

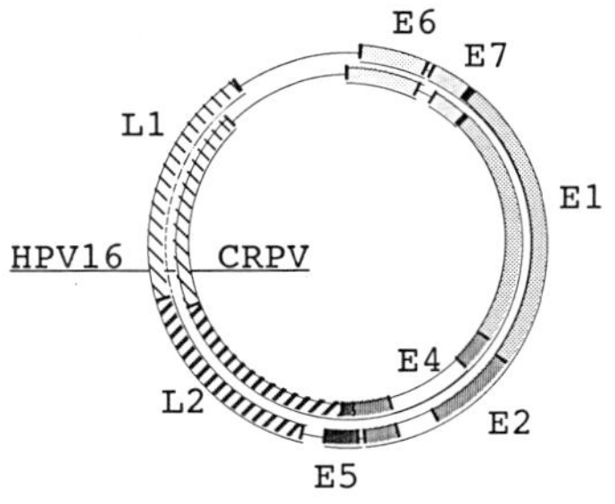

Predicted Protein	Amino Acid Similarity	Amino Acid Identity
E6	56	24
E7	38	34
E1	68	47
E2	51	34
E4	32	14
E5	42	16
L2	56	36
L1	72	54

Fig. 2. Homologies between CRPV and HPV 16 proteins.

lar clones of another papillomavirus, CRPV, can induce papillomas and carcinomas in rabbits, we felt that this approach was tenable for HPV in human tissue. Molecular clones of HPV genomes would offer an excellent alternative to whole virus because recombinant clones of many HPV types are available. In addition, numerous mutant HPV genes have already been characterized in vitro. In our model, newborn human foreskin is grafted orthotopically, i.e. from skin to skin, onto the backs of severe combined immunodeficient (scid) mice. Scid mice readily accept human xenografts and have previously been used to develop chimeric models of human disease [101]. The xenografts are inoculated with HPV DNA and monitored for the development of neoplasia. We have been able to induce exophytic papilloma with distinct histologic features of a viral lesion, including acanthosis, parakeratosis, and koilocytosis [J.L. Brandsma et al., unpubl. data], indicating that the approach will be successful.

Chimeric Papillomaviruses

Since all papillomaviruses have the same number and order of early and late open reading frames [H. Pfister, this issue], it may be possible to swap a gene or protein-coding domain of an animal papillomavirus for its human counterpart, and to inoculate the chimeric genome into animal skin to induce lesions that are caused, in part, by the human gene. Such a model could be used to evaluate the role of the HPV protein in the pathogenesis of disease. This approach has been successfully used with the papovaviruses JC and SV40 in transgenic mice [102].

The obvious model of choice for such studies is the CRPV-rabbit system, because it is currently the only one in which recombinant viral genomes readily induce disease. Significant homologies exist between CRPV genes and those of HPV 16 (fig. 2) [64a]. It is not known which viral genes and regulatory elements control species and tissue specificity, but the presence of conserved regions indicates that some genetic elements might be exchanged without hindering biological activity. We have generated CRPV/HPV chimeric genomes containing portions of HPV E7 genes in place of the CRPV counterpart. The only chimeras that have induced rabbit papillomas contained small substitutions of amino terminal E7 codons from HPV 1a or HPV 16, and the efficiency of papilloma induction was substantially reduced as compared to wild-type CRPV DNA [J.L. Brandsma et al. unpubl. data]. Nonetheless, these results indicate that the chimeric virus approach holds promise and also that the E7 gene encodes species-specific properties.

In summary, a complete understanding of papillomavirus gene functions in the etiology of human cancer can only be obtained in vivo. Considerable progress has been made in developing animal models of HPV-associated oncogenesis, using a variety of homologous, heterologous and chimeric systems. Further study of these complementary models will lead to a more refined understanding of the viral, cellular, and environmental factors that cooperate during oncogenesis.

Acknowledgments

I am grateful to Robert O. Jacoby and Daniel DiMaio for critical review of the manuscript. I would like to thank Lutz Gissmann and Douglas Hanahan for preprints.

References

1 zur Hausen H: Human papillomaviruses in the pathogenesis of anogenital cancer. Virology 1991;184:9–13.

2 Kraus SJ, Stone KM: Management of genital infection caused by human papillomavirus. Rev Infect Dis 1990;12(suppl 6):620–632.

3 Sousa R, Dostatni N, Yaniv M: Control of papillomavirus gene expression. Biochim Biophys Acta 1990;1032:19–37.

4 Lambert PF: Papillomavirus DNA replication. J Virol 1991;65:3417–3420.

5 McBride AA, Romanczuk H, Howley PM: The papillomavirus E2 regulatory proteins J Biol Chem 1991;266:18411–18414.

6 Munger K, Scheffner M, Huibregtse JM, Howley PM: Interactions of HPV E6 and E7 oncoproteins with tumour suppressor gene products. Cancer Surv 1992;12:197–217.

7 DiPaolo JA, Popescu NC, Alvarez L, Woodworth CD: Cellular and molecular alterations in human epithelial cells transformed by recombinant human papillomavirus DNA. Crit Rev Oncog 1993;4:337–360.

8 World Health Organization: Laboratory Techniques in Rabies, ed 2. Geneva, WHO, 1966.

9 Dollard SC, Wilson JL, Demeter LM, Bonnez W, Reichman RC, Broker TR, Chow LT: Production of human papillomavirus and modulation of the infectious program in epithelial raft cultures. Genes Dev 1992;6:1131–1142.

10 Meyers C, Frattini G, Hudson JB, Laimins LA: Biosynthesis of human papillomavirus from a continuous cell line upon epithelial differentiation. Science 1992;257:971–973.

11 Morey AL, Patou G, Myint S, Fleming KA: In vitro culture for the detection of infectious human parvovirus B19 and B19-specific antibodies using foetal haematopoietic precursor cells. J Gen Virol 1992;73:3312–3317.

12 Mitao M, Nagai N, Levine RU, Silverstein SJ, Crum CP: Human papillomavirus type 16 infection: A morphological spectrum with evidence for late gene expression. Int J Gynecol Pathol 1986;5:287–296.

13 Kreider JW, Howett MK, Leure-Dupree AE, Zaino RJ, Weber JA: Laboratory production in vivo of infectious human papillomavirus type 11. J Virol 1987;61:590–593.

14 Ito Y: Tumorigenesis with phenolic extracts of wild and domestic rabbit papilloma tissue and of partially purified Shope papillomavirus. Fed Proc 1961;20:438.

15 Ito Y, Evans CA: Induction of tumors in domestic rabbits with nucleic acid preparations from partially purified Shope papilloma virus and from extracts of the papillomas of domestic and cottontail rabbits. J Exp Med 1961;114: 485–500.

16 Ito Y: Studies on subviral tumorigenesis: Carcinoma derived from nucleic acid-induced papillomas of rabbit skin. Acta Unio Int Cancrum 1963;19:280–283.

17 Nasseri M, Wettstein FO: A variant of the CRPV DNA is preferentially maintained as a plasmid in NIH 3T3 cells and characterization of its transcripts in nude mouse tumors. Virology 1987;161:541–548.

18 Brandsma JL, Yang Z, Barthold SW, Johnson EA: Use of a rapid, efficient inoculation method to induce papillomas by cottontail rabbit papillomavirus DNA shows that the E7 gene is required. Proc Natl Acad Sci USA 1991;88: 4816–4820.

19 Georges E, Pehau-Arnaudet G, Orth G: Molecular and biological characterization of cottontail rabbit papillomavirus variant DNA sequences integrated in the VX7 carcinoma. Virology 1992;186:750–759.

20 Rowson KEK, Mahy BWJ: Human papova (wart) virus. Bacteriol Rev 1967;31:110.

21 Koller D, Olson C: Attempted transmission of warts from man to selected hosts. J Invest Dermatol 1972;58:366–368.

22 Sundberg JP: Animal models for papillomavirus research. Contrib Oncol 1987;24:11–38.

23 Amtmann E, Volm M, Wayss K: Tumour induction in the rodent *Mastromys natalensis* by activation of endogenous papilloma virus genomes. Nature 1984;308:291–292.

24 O'Banion MK, Reichmann ME, Sundberg JP: Cloning and characterization of a papillomavirus associated with papillomas and carcinomas in the European harvest mouse *(Micromys minutus)*. J Virol 1988;62:226–233.

25 Siegsmund M, Wayss K, Amtmann E: Activation of latent papillomavirus genomes by chronic mechanical irritation. J Gen Virol 1991;72:2787–2789.

26 Wettstein FO, Stevens JG: Distribution and state of viral nucleic acid in tumors induced by Shope papillomavirus; in Essex M, zur Hausen H (eds): Viruses in Naturally Occurring Cancers. Cold Spring Harbor, Cold Spring Harbor Laboratory, 1980, pp 301–307.

27 Watts SL, Ostrow RS, Phelps WC, Prince JT, Faras AJ: Free cottontail rabbit papillomavirus DNA persists in warts and carcinomas of infected rabbits and in cells in culture transformed with virus or viral DNA. Virology 1983;125:127–138.

28 Giri I, Danos O, Yaniv M: Genomic structure of the cottontail rabbit (Shope) papillomavirus. Proc Natl Acad Sci USA 1985;82:1580–1584.

29 O'Banion MK, Cialkowski ME, Reichmann ME, Sundberg JP: Cloning and molecular characterization of an oral papillomavirus of domestic rabbits. Virology 1988;162:221–231.

30 Howley PM, Law MF, Heilman C, Engel L, Alonso MC, Lancaster WD, Israel MA, Lowy DR: Molecular characterization of papilloma virus genomes; in Essex M, Todaro G, zur Hausen H (eds): Viruses in Naturally Occurring Cancers. Cold Spring Harbor Conf Cell Prolif 1980;7:233–247.

31 Law MF, Lowy DR, Dvoretzky I, Howley PM: Mouse cells transformed by bovine papillomavirus contain only extrachromosomal viral DNA sequences. Proc Natl Acad USA 1981;78: 2727–2731.

32 Patel KR, Smith KT, Campo MS: The nucleotide sequence and genome organization of bovine papillomavirus type 4. J Gen Virol 1987; 68:2117–2118.

33 O'Banion MK, Sundberg JP, Reichmann ME: Equine papillomavirus: Partial characterization and presence in common equine skin tumors. J Cell Biochem 1985;9:73.

34 Sundberg JP, O'Banion MK, Schmidt-Didier E, Reichmann ME: Cloning and characterization of a canine oral papillomavirus. Am J Vet Res 1986;47:1142–1144.

35 Kloster BE, Manias DA, Ostrow RS, Shaver MK, McPherson SW, Rangen SR, Uno H, Faras AJ: Molecular cloning and characterization of the DNA of two papillomaviruses from monkeys. Virology 1988;166:30–40.

36 Reszka AA, Sundberg JP, Reichmann ME: In vitro transformation and molecular characterization of Colobus monkey venereal papillomavirus DNA. Virology 1991;181:787–792.

37 Groff DE, Lancaster WD: Molecular cloning and nucleotide sequence of deer papillomavirus. J Virol 1985;56:85–91.

38 Stenlund A, Moreno-Lopez J, Ahola H, Pettersson U: European elk papillomavirus: Characterization of the genome, induction of tumors in animals, and transformation in vitro. J Virol 1983;48:370–376.

39 Moreno-Lopez J, Ahola H, Eriksson A, Bergman P, Petterson U: Reindeer papillomavirus transforming properties correlate with a highly conserved E5 region. J Virol 1987;61:3394–3400.

40 Amtmann E, Wayss K: Papillomaviruses and carcinogenic progression. II: The *Mastomys natalensis* papillomavirus; in Salzman NP, Howley PM (eds): The Papovaviridae. New York, Plenum Press, 1987, vol 2: The Papillomaviruses, pp 187–198.

41 Reeve VE, Greenoak GE, Canfield PJ, Boehm-Wilcox C, Tilbrook PA, Kulski JK, Gallagher CH: Enhancement of u.v.-induced skin carcinogenesis in the hairless mouse by inoculation with cell-free extracts of skin tumours. Immunol Cell Biol 1989;67:421–427.

42 Tilbrook PA, Greenoak GE, Reeve VE, Canfield PJ, Gissmann L, Gallagher CH, Kulski JK: Identification of papillomaviral DNA sequences in hairless mouse tumours induced by ultraviolet irradiation. J Gen Virol 1989;70: 1005–1009.

43 Sundberg JP, O'Banion MK, Shima A, Knupp C, Reichmann ME: Papillomas and carcinomas associated with a papillomavirus in European harvest mice *(Micromys minutus)*. Vet Pathol 1988;25:356–361.

44 Sundberg JP, O'Banion MK, Reichmann ME: Mouse papillomavirus: Pathology and characterization of the virus; in Steinberg BM, Brandsma JL, Taichman LB (eds): Cancer Cells 5/Papillomaviruses. Cold Spring Harbor, Cold Spring Harbor Laboratory, 1987, pp. 373–379.
45 9 CFR 71.3 (Code of Federal Regulations): Interstate movement of diseased animals and poultry generally prohibited. Animals and Animal Products. Revised as of January 1, 1994.
46 Fujimoto Y, Olson C: The fine structure of the bovine wart. Pathol Vet 1966;3:659–684.
47 DiMaio D: Transforming activity of bovine and human papillomaviruses in cultured cells. Adv Cancer Res 1991;56:133–159.
48 Campo MS: Papillomas and cancer in cattle. Cancer Surv 1987;6:39–54.
49 Shope RE: Infectious papillomatosis of rabbits. J Exp Dis 1933;58:607–624.
50 Kreider JW, Bartlett GL: The Shope papilloma-carcinoma complex of rabbits: A model system of neoplastic progression and spontaneous regression. Adv Cancer Res 1981;35:81–110.
51 Wettstein FO: Papillomaviruses and carcinogenic progression. I: CRPV (Shope) papillomavirus; in Salzman NP, Howley PM (eds): The Papillomaviruses. New York, Plenum, 1987, vol 2, pp 167–186.
52 Noyes WF, Mellors RC: Fluorescent antibody detection of the antigens of the Shope papilloma virus in papillomas of the wild and domestic rabbit. J Exp Med 1957;106:555–562.
53 McMichael H: Inhibition by methylprednisolone of regression of the Shope rabbit papilloma. J Natl Cancer Inst 1967;39:55–65.
54 Okabayashi M, Angel MG, Christensen ND, Kreider JW: Morphometric analysis and identification of infiltrating leucocytes in regressing and progressing Shope rabbit papillomas. Int J Cancer 1991;49:919–923.
55 Rous P, Beard JW: The progression to carcinoma of virus-induced rabbit papillomas (Shope). J Exp Med 1935;62:523–548.
56 Rous P, Kidd JG: The carcinogenic effect of a papilloma virus on the tarred skin of rabbits. I: Description of the phenomenon. J Exp Med 1938;67:399–428.
57 Rous P, Kidd JG: The activation, transforming and carcinogenic effects of the rabbit papilloma virus (Shope) upon implanted tar tumors. J Exp Med 1940;71:787–812.
58 Syverton JT: The pathogenesis of rabbit papilloma to carcinoma sequence. Ann NY Acad Sci 1952;54:1126–1140.
59 Kidd JG, Rous P: Cancers deriving from the virus papillomas of wild rabbits under natural conditions. J Exp Med 1940;71:469–494.
60 Han R, Breitburd F, Marche PN, Orth G: Linkage of regression and malignant conversion of rabbit viral papillomas to MHC class II genes. Nature 1992;356:66–68.
61 Wank R, Thomssen C: High risk of squamous cell carcinoma of the cervix for women with HLA-DQw3. Nature 1991;352:723–725.
62 Kreider JW, Balogh K, Olson RO, Martin JC: Treatment of latent rabbit and human papillomavirus infections with 9-(2-phosphonylmethoxy)ethylguanine (PMEG). Antiviral Res 1990; 14:51–58.
63 Ostrow RS, Forslund KM, McGlennen RC, Shaw DP, Schlievert PM, Ussery MA, Huggins JW, Faras AJ: Ribavirin mitigates wart growth in rabbits at early stages of infection with cottontail rabbit papillomavirus. Antiviral Res 1992;17:99–113.
64 Brandsma JL, Xiao W: Infectious Virus Replication in Papillomas Induced by Molecularly Cloned Cottontail Rabbit Papillomavirus DNA. J Virol 1993;67:567–571.
64a Wu X, Xiao W, Brandsma JL: Papilloma formation by cottontail rabbit papillomavirus requires E1 and E2 regulatory genes in addition to E6 and E7 transforming genes. J Virol 1994; 68:6097–6102.
65 Meyers C, Harry J, Lin YL, Wettstein FO: Identification of three transforming proteins encoded by cottontail rabbit papillomavirus. J Virol 1992;66:1655–1664.
66 Nasseri M, Meyers C, Wettstein FO: Genetic analysis of CRPV pathogenesis: The L1 open reading frame is dispensable for cellular transformation but is required for papilloma formation. Virology 1989;170:321–325.
67 Brandsma JL, Yang Z, DiMaio D, Barthold SW, Johnson EA, Xiao W: The putative E5 open reading frame of cottontail rabbit papillomavirus is dispensable for papilloma formation in domestic rabbits. J Virol 1992;66:6204–6207.
68 Defeo-Jones D, Vuocolo GA, Haskell KM, Hanobik MG, Kiefer DM, McAvoy EM, Ivey-Hoyle M, Brandsma JL, Oliff A, Jones RE: Papillomavirus E7 protein binding to the retinoblastoma protein is not required for viral induction of warts. J Virol 1993;67:716–725.
69 Haskell KM, Vuocolo GA, Defeo-Jones D, Jones RE, Ivey-Hoyle M: Comparison of the binding of the human papillomavirus type 16 and cottontail rabbit papillomavirus E7 proteins to the retinoblastoma gene product. J Gen Virol 1993;74:115–119.
70 Sundberg JP, Junge RE, El Shazly MO: Oral papillomatosis in New Zealand white rabbits. Am J Vet Res 1985;46:664–668.
71 Sundberg JP, Schmidt E, O'Banion K, Olson C: Squamous cell carcinoma associated with canine oral papillomavirus. J Cell Biochem 1975; 9:75.
72 Bregman CL, Hirth RS, Sundberg JP, Christensen EF: Cutaneous neoplasms in dogs associated with canine oral papillomavirus vaccine. Vet Pathol 1987;24:477–487.
73 O'Banion MK, Sundberg JP, Shima AL, Reichmann ME: Venereal papilloma and papillomavirus in a colobus monkey *(Colobus guereza)*. Intervirology 1987;28:232–237.
74 Sundberg JP, Shima AL, Adkison DL: Oral papillomavirus infection in a pygmy chimpanzee *(Pan paniscus)*. J Vet Diagn Invest 1992;4: 70–74.
75 Ostrow RS, McGlennen RC, Shaver MK, Kloster BE, Houser D, Faras AJ: A rhesus monkey model for sexual transmission of a papillomavirus isolated from a squamous cell carcinoma. Proc Natl Acad Sci USA 1990;87:8170–8174.
76 Ostrow RS, LaBresh KV, Faras AJ: Characterization of the complete RhPV 1 genomic sequence and an integration locus from a metastatic tumor. Virology 1991;181:424–429.
77 Schneider JF, McGlennen RC, LaBresh KV, Ostrow RS, Faras AJ: Rhesus papillomavirus type 1 cooperates with activated *ras* in transforming primary epithelial rat cells independent of dexamethasone. J Virol 1991;65:3354–3358.
78 Ostrow RS, Liu Z, Schneider JF, McGlennen RC, Forslund K, Faras AJ: The products of the E5, E6, or E7 open frames of RhPV 1 can individually transform NIH 3T3 cells or in cotransfections with activated *ras* can transform primary rodent epithelial cells. Virology 1993; 196:861–867.
79 Matsukura T, Kanda T, Furuno A, Yoshikawa H, Kawana T, Yoshiike K: Cloning of monomeric human papillomavirus type 16 DNA integrated within cell DNA from a cervical carcinoma. J Virol 1986;58:979–982.
80 Baker CC, Phelps WC, Lindgren V, Braun MJ, Gonda MA, Howley PM: Structural and transcriptional analysis of human papillomavirus type 16 sequences in cervical carcinoma cell lines. J Virol 1987;61:962–971.
81 Kirnbauer R, Taub J, Greenstone H, Roden R, Durst M, Gissmann L, Lowy DR, Shiller JT: Efficient self-assembly of human papillomavirus type 16 L1 and L1-L2 into virus-like particles. J Virol 1993;67:6929–6936.
82 Lacey M, Alpert S, Hanahan D: Bovine papillomavirus genome elicits skin tumours in transgenic mice. Nature 1986;322:609–612.
83 Hanahan D, Wetzel E, Skowronski J, Sippola-Thiele M, Lindgren V, Howley P: Tumorigenic latency and separable stages during fibrosarcoma development in transgenic mice carrying papillomavirus genomes. Int Symp Princess Takamatsu Cancer Res Fund 1989;20:289–296.
84 Sippola-Thiele M, Hanahan D, Howley PM: Cell-heritable stages of tumor progression in transgenic mice harboring the bovine papillomavirus type 1 genome. Mol Cell Biol 1989;9: 925–934.
85 Lindgren V, Sippola-Thiele M, Skowronski J, Wetzel E, Howley PM, Hanahan D: Specific chromosomal abnormalities characterize fibrosarcomas of bovine papillomavirus type 1 transgenic mice. Proc Natl Acad Sci USA 1989; 86:5025–5029.
86 Peng X, Olson RO, Chistian CB, Lang CM, Kreider JW: Papillomas and carcinomas in transgenic rabbits carrying Ej-*ras* DNA and cottontail rabbit papillomavirus DNA. J Virol 1993;67:1698–1701.
87 Tinsley JM, Fisher C, Searle PF: Abnormalities of epidermal differentiation associated with expression of the human papillomavirus type 1 early region in transgenic mice. J Gen Virol 1992;73:1251–1260.

88 Kondoh G, Murata Y, Aozasa K, Yutsudo M, Hakura A: Very high incidence of germ cell tumorigenesis (seminomagenesis) in human papillomavirus type 16 transgenic mice. J Virol 1991;65:3335–3339.
89 Arbeit JM, Munger K, Howley PM, Hanahan D: Neuroepithelial carcinomas in mice transgenic with human papillomavirus type 16 E6/E7 ORFs. Am J Pathol 1993;142:1187–1197.
90 Griep AE, Herber R, Jeon S, Lohse JK, Dubielzig RR, Lambert PF: Tumorigenicity by human papillomavirus type 16 E6 and E7 in transgenic mice correlates with alterations in epithelial cell growth and differentiation. J Virol 1993;67:1373–1384.
91 Lambert PF, Pan H, Pitot HC, Liem A, Jackson M, Griep AE: Epidermal cancer associated wtih expresion of human papillomavirus type 16 E6 and E7 oncogenes in the skin of transgenic mice. Proc Natl Acad Sci USA 1993;90: 5583–5587.
92 Arbeit JM, Munger K, Howley PM, Hanahan D: Progressive squamous epithelial neoplasia in K14-human papillomavirus type 16 transgenic mice. J Virol, in press.
93 Howett MK, Kreider JW, Cockley KD: Human xenografts: A model system for human papillomavirus infection. Intervirology 1990;31:109–115.
94 Kreider JW, Howett MK, Wolfe SA, Bartlett GL, Zaino RJ, Sedlacek T, Mortel R: Morphological transformation in vivo of human uterine cervix with papillomavirus from condylomata acuminata. Nature 1985;317:639–641.
95 Kreider JW, Howett MK, Lill NL, Bartlett GL, Zaino RJ, Sedlacek TV, Mortel R: In vivo transformation of human skin with human papillomavirus type 11 from condylomata acuminata. J Virol 1986;59:369–376.
96 Kreider JW, Howett MK, Stoler MH, Zaino RJ, Welsh P: Susceptibility of various human tissues to transformation in vivo with human papillomavirs type 11. Int J Cancer 1987;39: 459–465.
97 Brown DR, Chin MT, Strike DG: Identification of human papillomavirus type 11 E4 gene products in human tissue implants from athymic mice. Virology 1988;165:262–267.
98 Christensen ND, Kreider JW, Cladel NM, Patrick SD, Welsh PA: Monoclonal antibody-mediated neutralization of infectious human papillomavirus type 11. J Virol 1990;64: 5678–5681.
99 Christensen ND, Kreider JW, Shah KV, Rando RF: Detection of human serum antibodies that neutralize infectious human papillomavirus type 11 virions. J Gen Virol 1992;73: 1261–1267.
100 Sterling J, Stanley M, Gatward G, Minson T: Production of human papillomavirus type 16 virions in a keratinocyte cell line. J Virol 1990;64:6305–6307.
101 Hendrickson EA: The SCID mouse: Relevance as an animal model system for studying human disease. Am J Pathol 1993;143:1511–1522.
102 Ressetar HG, Prakash O, Frisque RJ, Webster HD, Re RN, Stoner GL: Expression of viral T-antigen in pathological tissues from transgenic mice carrying JC-SV40 chimeric DNAs. Mol Chem Neuropathol 1993;20:59–79.

Intervirology 1994;37:201–214

Achim Schneider

Department of Gynecology, University of Jena, Germany

Natural History of Genital Papillomavirus Infections

Key Words
Human papillomavirus
Natural history
Course of infection
Biologic potential
Transmission
Diagnosis
Immunity

Introduction

Human papillomaviruses (HPVs) are causally associated with the development of cervical cancer in otherwise healthy women [1] whereas HPV-associated skin cancers are mainly found in immunosuppressed individuals or patients suffering from epidermodysplasia verruciformis (EV) [2, 3]. The importance of certain HPV types for anogenital and cutaneous carcinogenesis has been well studied and established in the last decade. However, our knowledge regarding the natural history of HPV infections is still fragmentary. In this article, diagnosis, course of infection, transmission, and risk factors will be reviewed.

Definition and Diagnosis

Infection of the keratinocyte by HPV may result in different morphologic phenomena which can be classified as clinical, subclinical, and latent infections (fig. 1) [4]. Alternatively, clinical infections may be discriminated from cytological or molecular infections [5]. HPV infections should be distinguished from HPV-associated disease (premalignant or malignant neoplasia) since in addition to HPV other factors are needed to transform the keratinocyte. For categorization of premalignant disease of the lower genital tract several morphologic systems are in use (fig. 2). The extent of involvement of the epithelium by dyskaryotic cells and cytomorphologic features of the individual cell, where the presence of atypical mitotic figures indicates high-grade disease, are the most important features [6]. However it should be kept in mind that, though considered the gold standard for classification of intraepithelial neoplasia, histology is by no means a well-reproducible and accurate method, especially for the assessment of minor or low-grade disease: only poor agreement is observed between different experts [7].

In reviewing published data on HPV it is important to understand the validity of the various HPV detection systems in use: standardization of the different techniques is not yet available [8], making comparison of results generated by different laboratories difficult [9, 10]. For diagnosis of HPV infections, viral DNA, RNA, proteins, or antibodies may be detected. Detection of HPV DNA is the most commonly used test varying in sensitivity between a threshold limit of a total of 10 HPV copies for polymerase chain reaction (PCR)-based techniques and 10,000 HPV copies for conventional hybridization techniques such as Southern blot, dot blot, sandwich, and in situ hybridization [11, 12]. For epidemiological studies, PCR techniques are predominantly used, allowing identification of a wide range of HPV types [13, 14]. Since 72–91% of low-grade lesions, 90–100% of high-grade lesions [15–17], and 84–100% of invasive cancers [17–21] are HPV positive

Achim Schneider, MD, MPH
Department of Gynecology
Friedrich Schiller University of Jena
Bachstrasse 18
D-07743 Jena (Germany)

0300–5526/94/0374–0201
$8.00/0

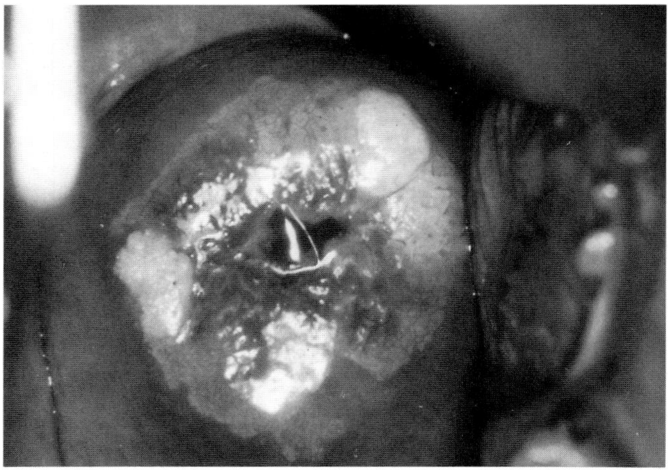

Fig. 1. Colpophotography showing an atypical transformation zone with condylomata acuminata (clinical HPV infection) surrounded by flat condyloma (subclinical HPV infection) and possibly the presence of HPV in the normal epithelium surrounding the lesion (latent HPV infection).

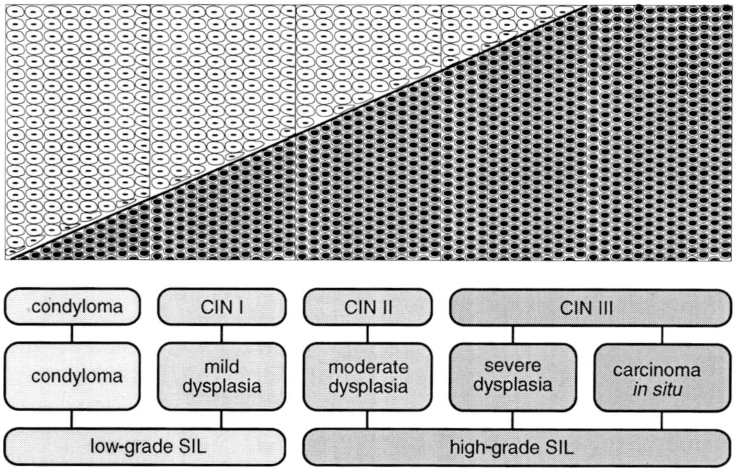

Fig. 2. Histologic classification systems used for cervical precancerous lesions. The cervical intraepithelial neoplasia (CIN) classification [179] is also applied to other areas of the lower genital tract and is used for the vagina (vaginal intraepithelial neoplasia, VAIN I–III), vulva (vulvar intraepithelial neoplasia, VIN I–III), anus (anal intraepithelial neoplasia, AIN I–III) and penis (penile intraepithelial neoplasia, PIN I–III). The dysplasia-carcinoma in situ classification has the longest tradition and is still widely used, especially in Europe [180]. The Bethesda system [181] was initially designed for reporting cytological findings but is meanwhile also accepted and used by pathologists.

with PCR, the sensitivity of this technique seems satisfactory. The HPV-type specificity of PCR results equals that found by Southern blot hybridization, the gold standard for HPV typing [22]. For epidemiological studies, it also seems useful to group various HPV types into 'low-risk' and 'high-risk' HPV categories, a division which is supported by phylogenetic trees which classify HPVs into specific groups encompassing the known tissue tropism and oncogenic potential of each HPV type (table 1) [23].

Detection of serum antibodies has the advantage of being independent of proper collection of HPV-containing cells or tissues. For detection of HPV-type-specific antibodies, bacterial fusion proteins from some early and all late open reading frames (ORFs) of HPV 6, HPV 11, HPV 16, and HPV 18 [24–31] and immunoreactive synthetic peptides designated for HPV 6 L2, HPV 11 L2, HPV 16 E2, L1, L2, and HPV 18 E7, L2 [31–42] have been used to investigate the meaning of humoral immune response for the natural history of genital HPV infections. Antibodies against the E6 and E7 proteins of HPV 16 as detected by radioimmunoprecipitation of in vitro translated proteins are detected in 72% of patients with invasive cervical cancer compared with 5.8% positive sera in controls [43]. This significant association was, however, not confirmed in serum from patients with high-grade squamous intraepithelial lesions (SILs) [43, 44]. Thus, HPV antibody testing with the currently available techniques may only be useful in patients with invasive disease. Since intact virions present conformational epitopes, they should be more suitable for detection of antibodies compared with bacterial fusion proteins or synthetic peptides [45]. HPV 11 virions generated in the nude mouse xenograft system reacted significantly more frequently with serum from patients with condylomata acuminata and recurrent respiratory papillomatosis (RRP) compared to controls [46, 47]. Expression of the

Table 1. Classification of HPV

	Epidemiology	Phylogeny
Low risk	HPV 6, 11, 42, 43, 44	HPV 6, 11, 40, 42, 43, 44, 57
Intermediate risk	HPV 31, 33, 35, 51, 52, 58	
High risk	HPV 16, 18, 45, 56	HPV 16, 18, 30, 31, 33, 35, 39, 45, 51, 52, 56, 58, 59, 66
Unknown risk		HPV 61, X

The epidemiologic classification of genital HPV types is derived from a cross-sectional study examining 2,627 biopsies of different histopathologic grade from the lower genital tract [22]. Taxonomic classification of genital HPV types is according to phylogenetic trees based on nucleotide sequence [23].

early genes of high-risk HPVs in eukaryotic expression systems and assembly of HPV-specific proteins in virus-like particles may therefore yield a higher detection rate in HPV-infected patients [48; I. Nindl, pers. commun.].

Macromorphology is not a reliable method for defining HPV-type-specific lesions, although high-risk HPV types are prevalent in papular and especially macular penile condyloma-like lesions [49], but pigmentation does not correlate with the presence of high-grade penile intraepithelial neoplasia and the presence of HPV 16 [50]. Histomorphological changes may be specific for certain HPV types as shown for cutaneous warts [51] and a subset of genital lesions [52]. Although the Bethesda system (fig. 2) may have the advantage of better reproducibility compared to other systems, grading of lesions correlates only with the presence of the high-risk HPV type 16 but not with high-risk HPV types 18, 31, 35, and 39 [53]. As also shown for penile lesions, no significant difference was found in the prevalence of HPV infection between patients (n = 78) with acetowhite-positive and -negative reactions in the genital mucosa using PCR testing [54]. Thus, morphologic methods are nonspecific for the diagnosis of nonclinical disease and less than 30% of tissues suspicious for subclinical HPV infection contain HPV even when using the highly sensitive PCR technique [55–57]. Therefore, application of virological diagnostic tests is obligatory for valid epidemiological HPV research.

Course of Infection

HPV infects basal cells of the differentiating epithelium. The time course of molecular biologic events has been studied by in situ hybridization on human xenografts in nude mice infected with HPV 11 which is the natural cause of condylomata acuminata [58]. Four weeks after initial infection, transcription of the early ORFs is initiated. Viral replication and transcription of early and late ORFs reach their maximum between week 6 and 8. After 10–12 weeks, condylomata are established. The highest amount of viral DNA and viral antigen production takes place in the most differentiated superficial cells [59]. The sequence of events in subclinically and latently infected tissue is not known due to the lack of sensitivity of the currently available detection systems [60].

The regression rate of condylomata acuminata varies between 0 and 69% in placebo-treated patients [61–64]. However, in up to 67%, disease may recurr and 45% of patients with condylomata acuminata stay latently infected after treatment as shown by Southern blot analysis [65]. In patients with RRP (n = 12) analysis by dot blot hybridization showed 4 of 10 nondiseased sites positive for HPV 6/11 [66].

The time course of subclinical and latent disease has been studied with the limitation that examinations did not exclude reinfection by sexual partners. Follow-up of up to 27 months shows that between 50 and 66% of patients are HPV positive on at least one occasion in their cervical smear using Southern blot hybridization, dot blot hybridization, or PCR-based HPV detection [67–69]. Between 30 and 50% of women remain intermittently or consistently positive depending on the sensitivity of the HPV-testing system used and up to 40% of the subjects have new or different HPV types compared to the original HPV type [68, 69]. Thus, current knowledge about the time course of HPV infections can be schematically summarized as shown in figure 3.

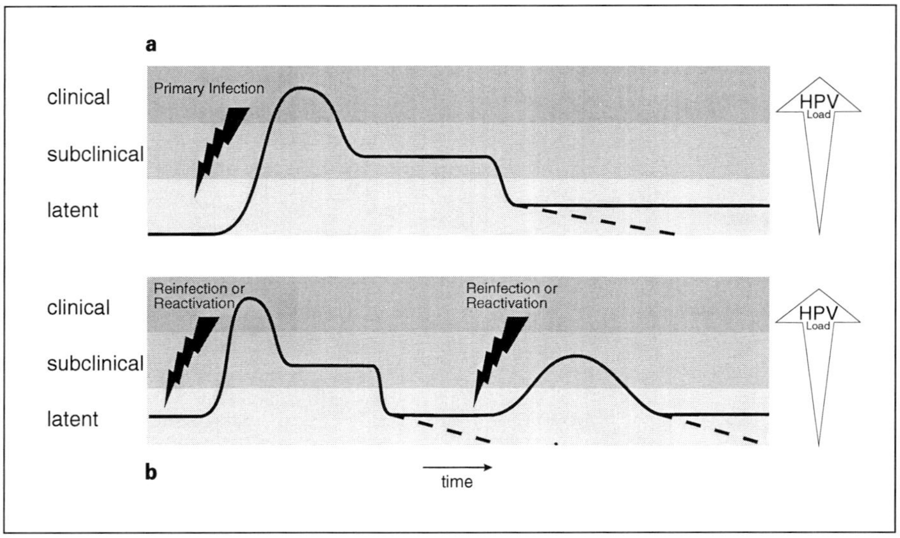

Fig. 3. a Clinical disease may evolve after primary infection. These lesions may regress spontaneously or result in persistent latent infection. Whether the first infection can primarily be subclinical or latent is questionable. HPV replication and viral load increase linearly from latent to subclinical and clinical infection. **b** After latent HPV infection is established, reactivation or reinfection may occur. Reactivation or reinfection may result in subclinical or clinical disease. Clinical disease may regress via subclinical disease or go directly to latent HPV infection.

Infections with high-risk HPVs are associated with a relative risk of between 8 and 11 for development of SILs [70; M. Schiffman, pers. commun.]. In addition to viral type, persistence of positivity for high-risk HPV types in cervical and anal smears and high viral load are strongly associated with progression to anogenital intraepithelial neoplasia [68; M. Schiffman, et al., pers. commun.]. Transcription of the E6-E7 region of high-risk HPVs could be an additional prerequisite and possible indicator for progression [71, 72], and mutations of transforming HPV genes as shown for E7 may be important for variations of biological behavior of high-risk HPVs in different geographic regions [73]. The severity of SILs seems to vary for different cohorts: in women with a low prevalence of other risk factors for SILs, low-grade lesions are the main diagnosis, in contrast to patients from a sexually transmitted diseases (STD) clinic in which incident high-grade SILs appeared shortly after detection of HPV 16 or 18 DNA among women previously negative by cytology [74; M. Schiffman, pers. commun.]. Thus, different modes of progression may exist, and this is confirmed by a recent descriptive study analyzing the mode ages of women with SILs [75]: the age distribution in patients with SILs is not normal but skewed from a mean age of 29 years to a mode age of 22 years for low-grade SILs and in patients with carcinoma in situ from a mean age of 36 years to mode ages of 24, 30, and 32 years. Thus, certain cohorts of patients may progress much quicker (within 2 years) than others (within 8 to 10 years) which again may depend on different risk factors.

Regression of HPV infections and HPV-associated neoplasia depends mainly on a competent cellular immune response. In regressing skin warts, mononuclear cell inflammatory infiltrates containing macrophages, T lymphocytes, and natural killer (NK) cells are the major morphological feature [76–78]. This reaction has a biphasic course in the cottontail rabbit papillomavirus (CRPV) model: infiltrates consist of polymorphonuclear cells during the early stage and are replaced by predominantly perivascular cell infiltrates composed of mononuclear cells in the later stage of the local immune response [79]. Cellular immune impairment is associated with condyloma acuminatum in humans since T helper cells are depleted and the T4:T8 ratio is reversed in tissue and peripheral blood [80, 81]. The number of Langerhans cells is decreased in cervical-HPV-associated lesions [82–85], but these Langerhans cells show a higher expression of major histocompatibility complex (MHC) class II antigen [83]. Recognition of HPV-16-infected keratinocytes by NK cells depends on persistence of the lesion and cannot be corrected by incubation of peripheral blood mononuclear cells (PBMCs) with interferon-γ and/or interleukin-2, although these cytokines are decreased in condylomata [81, 86, 87]. Leukoregulin activation of NK cells, however, increases lymphocyte lysis of HPV-positive cervical cancer cells and HPV-immortalized human cervical epithelial cells [88]. Modification of the expression of MHC molecules [85, 89] may influence the ability of keratinocytes to function as antigen-presenting cells [90; C. Soler, pers. commun.]. MHC class I expression is down-regulated in cervical cancer and possibly in SILs, whereas

MHC class II is expressed in the majority of cervical cancers and SILs [89, 91, 92]. MHC class I down-regulation in cervical cancer is controlled posttranscriptionally and independently of HPV 16 E7 transcription and/or overexpression of c-*myc* [91]. In humans, the local immune response was tested by skin tests using in-vitro-translated viral fusion proteins: reactions against proteins of the HPV 16 L1 ORF were seen in 6 of 7 patients with SILs compared to no reaction in 10 controls [93]. A significant difference in skin reactivity against intact and/or denatured CRPV particles or in-vitro-translated capsid proteins was seen between regressor (10 of 11) and progressor (10 of 20) papillomas in domestic rabbits with recall reactions detectable from several months up to 2 years postregression [94]. Cytotoxic T cells with specific binding to HPV 16 E6 and E7 oncogene sequences have been established and show potential use for treatment of HPV-associated disease [W.M. Kast, pers. commun.].

The humoral immune response can protect against papillomavirus infections in animal models by neutralizing antibodies against late (capsid) viral proteins [95, 96; M.S. Campo, pers. commun.]. The full-length, nondenatured L1 product seems necessary for induction of neutralizing antibodies, protection from papilloma development, and prevention of latent infection as shown in the CRPV system [97].

Transmission

Sexual Transmission

The incubation period for condylomata acuminata varies between 3 weeks and 8 months [98, 99]. For subclinical and latent HPV infections no data about interval between exposure and establishment of infection are available. Up to 70% of male partners of women with cervical HPV infections and intraepithelial neoplasia are diagnosed with subclinical HPV infections or HPV-associated disease [100, 101]. However, for HPV-16-positive women only 47% of lesions diagnosed in husbands also contained the same virus [102]. To complicate issues even more, only 50% of HPV-16-positive couples harbored identical HPV 16 variants. This, however, could also be due to the presence of different HPV 16 variants in the same lesion, as seen in 16% of the cases examined, and lack of sensitivity for detection of the HPV variant present in the partner [102]. Virus-containing lesions or cells may act as the main source for transmission but semen and/or the urethra may also act as a HPV reservoir since there is a significant difference in HPV positivity in female consorts of men with HPV-16-positive semen compared to men with HPV-16-negative semen [S. Kyo, pers. commun.]. Semen of 29 males with genital warts was positive by PCR in 83% and in 33% by Southern blot hybridization, compared with only 41% (43/104) by PCR and 0% (0/70) by Southern blot hybridization in males seen for infertility [103]. These findings were not confirmed in another study examining 17 male partners of women with HPV infection using dot blot hybridization, who were all HPV negative, as were 16 male controls [104]. In 65 men with multiple heterosexual partners, 17% of the distal urethral specimens were positive for HPV 6/11, 16, 18, or 33 by PCR [105]. Similar findings were reported in 64 male partners of women with genital HPV infection (n = 64) where 21% of the distal urethral specimens were positive for HPV 6/11, 16, or 18 by PCR, and 30% of urethral swabs from men with visible penile lesions and 18% from men without visible lesions were HPV positive [106]. Another study of 39 men with genital dermatoses found 25% of urethral swab specimens and 10% of urine specimens positive for HPV using PCR and Southern blot hybridization, but there was no difference in HPV detection between men with histological evidence of warts or neoplasia and men without such changes [107]. The rate of HPV positivity by PCR in urine specimens (38%) was almost equal to the prevalence in cervical specimens (49%) in 489 women examined in a colposcopy clinic: 65% of urine samples were HPV positive in patients with HPV-positive cervical specimens versus 84% HPV-positive cervical swabs in patients with a HPV-positive urine specimen [108]. A positive HPV urine test due to contamination from external lesions is rather unlikely since 8% of 138 presumably healthy military conscripts had HPV-DNA-positive urethra samples and 5% had HPV-DNA-positive urine samples [108]. However, in another recent study, 73 patients attending an STD clinic, including 14 patients with genital warts, were HPV DNA negative by PCR in their urinary sediment [109]. Infection of the bladder mucosa could explain positive urine results, and the presence of HPV DNA in transitional cell carcinoma of the bladder is indirect evidence that latent or subclinical HPV infection may exist in this organ: high-risk HPVs were detected in 31% of 90 bladder cancers with multiple HPV types present in 61% of these tumors [110]. Though there is evidence that the urethra, semen, and/or bladder may act as a HPV reservoir, contamination of samples by external lesions cannot be excluded. Presence of the virus in the reservoir tissues has to be proven directly, especially since nonsquamous epithelium was shown to be nonsusceptible for propagation of HPV 11 [111].

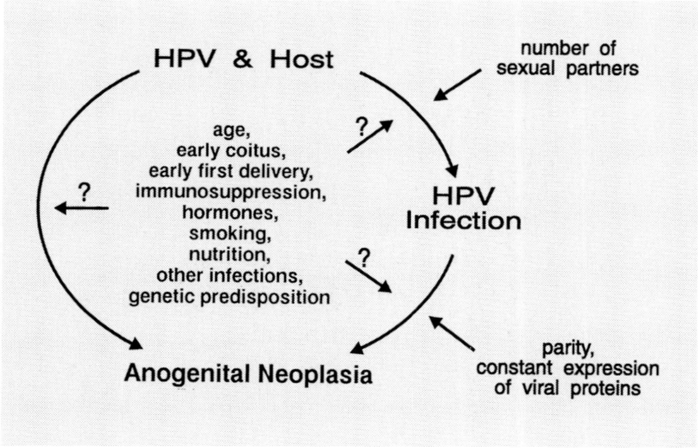

Fig. 4. Risk factors for genital HPV infections and HPV-associated disease and their assumed point of action (modified after a poster by H. Bauer and M. Manos).

Nonsexual Transmission

RRP is a well-recognized entity and the result of perinatal transmission of HPV [112]. The extent and mode of transmission to the oronasopharyngeal and genital area of the newborn are not clear since HPV detection rates in nasopharyngeal aspirates from newborns vary between 2.8 and 33% [113–115], rates similar to findings in oral cavity scrapings by type-specific PCR of 21 preschool children (24% for HPV 6 and 19% for HPV 16) and 35 adult men (17% for HPV 6 and 23% for HPV 16) [27]. Vertical transmission to the fetal lower genital tract seems unlikely since a recent study on neonatal foreskins and cervical specimens from autopsy specimens of 52 males and 46 females, of which 80 had died in utero, 3 during labor and 15 following delivery, was negative for HPV by PCR; cytological evidence of HPV infection was found in 8.6% of mothers prior to delivery [116]. The oral cavity may act as a reservoir for genital HPVs: high-risk HPV types may play an important role for induction of malignancy in the upper respiratory and upper gastrointestinal tract as shown for tonsillar carcinomas with HPV 16 and HPV 33 [117], and for 23% of 363 squamous cell cancers from the esophagus positive for HPV by in situ hybridization [118]; HPV 16 was the most prevalent type, occurring in 18.8% of the 85 HPV-positive specimens. In the skin, HPV types 16 and 35 are also present in intraepithelial or invasive neoplasia of the periungual area [119, 120], and fomites were found to be positive for HPV DNA even after sterilization [121]. The meaning of these findings for transmission of genital HPV infections is not understood.

Risk Factors for Detection of Genital HPV Infection

The presence of intraepithelial or invasive disease is the strongest risk factor for harboring HPV in the lower genital tract since almost 100% of such lesions are HPV positive [22]. HPV infection precedes the development of SILs (see 'Course of Infection'), but genital smears from women with a history of SILs or cervical cancer are also about 3-fold more frequently positive for HPV compared to controls [17, 122]. The prevalence of HPV infections in the general population is much lower than in patients with HPV-associated neoplasia and depends on various risk factors which have been investigated with respect to their association with the development of anogenital neoplasia [123]. Recent studies have also addressed the question whether such risk factors increase the risk for HPV infection, for HPV-associated neoplasia, or for both (fig. 4).

Sexual Behaviour

The number of sexual partners increases the risk for HPV positivity in the lower genital tract [124–126], and this association is independent of other risk factors such as age, race or use of oral contraceptives [127]. The relative risk for acquisition of HPV with each new partner was estimated to be 3.1 [L.A. Koutsky, pers. commun.]. Early coitus was shown to be an independent risk factor for cervical cancer with an odds ratio (OR) of 4.3 for age <16 years versus 24+ years in a case-control study in Colombia and Spain where HPV status was measured by consensus-PCR [128]. However, the number of sexual partners and early coitus may be mainly responsible for the acquisiton of genital HPV infection and not for the transformation of HPV infection to HPV-associated neoplasia since relative risks seen for an increasing number of sexual partners and young age at coitus in patients with SILs were neutralized after adjustment for HPV (fig. 4) [129].

Immunosuppression

An increased risk for genital warts and cervical intraepithelial neoplasia (CIN) has been described in renal allograft recipients [130, 131]. Latent and subclinical HPV infections with HPV 16/18 were found in 27% of women with renal allografts (n = 49), compared to only 6% in controls [132]. Therapeutic immunosuppression in renal transplant patients induces characteristic changes in their HPV antibody profile: whereas IgG antibodies against HPV 16 L1 and L2 peptides dropped significantly after transplantation, IgA activity against the E2 region increased significantly, suggesting reactivation of latent virus [133].

Human Immunodeficiency Virus (HIV)

HIV-positive patients have a higher rate of HPV infections and, possibly, HPV-associated neoplasia than controls. In HIV-positive cytologically negative women, 4 out of 10 with a CD4 count <200/mm^3 were HPV positive by dot blot compared to none of 22 women with a CD4 count ≥200/mm^3 [134]. Symptomatic HIV-infected women (n = 33) showed a significantly higher HPV prevalence (70%) in cervicovaginal lavages by Southern blot hybridization than HIV-positive asymptomatic women (4 of 18; 22%), or HIV-negative high-risk women (10 of 45; 22%) [135]. In this study, the risk for SILs was significantly higher [OR 12; 95% confidence interval (CI) 1.3–108] in symptomatic HIV-positive women compared with controls. This was not confirmed by others: in East African women, HPV 16 and 18 were 2.2-fold more frequently detected in HIV-positive asymptomatic women by Southern blot hybridization and PCR but no association was seen with the rate of abnormal Pap smears [20]. 32% of HIV-positive prostitutes were HPV positive, compared to 7% of HIV-negative prostitutes, but the rate of abnormal cervical cytology did not differ between the two groups [136].

In men, 31–53% of anal swabs of HIV-positive individuals were HPV positive by dot blot hybridization or PCR compared to only 8–29% HPV positives in HIV-negative individuals (n = 119) [137, 138]. A T4/T8 ratio of <0.4 was associated with a significantly higher HPV positivity rate of 35.3% compared with only 7.3% in men with a ratio of >1.0 [139]. The presence of HPV is strongly associated with cytological abnormalities in anal smears of homosexual men [137, 139], and persistent infection with HPV 16 or 18 seems the most important risk factor for the development of anal SILs in HIV-positive men [C. Critchlow, pers. commun.]. A history of intravenous drug use, hairy leukoplakia, or CD4 counts less than 200/mm^3 are additional risk factors for disease progression [J.M. Palefsky, pers. commun.].

An association between HIV and HPV is currently explained by decreased cellular immunity in HIV-infected individuals, which increases the risk for the acquisition of HPV infection and/or the reactivation of latent HPV infection. Synergism between these viruses may be explained by transactivation of the long control region (LCR) by the HIV-1 *tat* protein which increases HPV 18 E7 expression in HeLa cells [140]. This observation needs further clarification, especially since HPV and HIV do not share target cells.

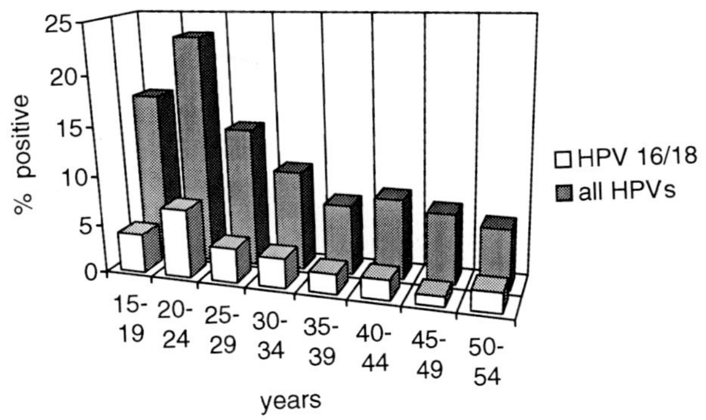

Fig. 5. Age dependency of HPV detection in genital smears of 4,144 female hospital patients detected by general primer PCR [144].

Other Infectious Agents

Infections with herpes simplex virus 2 or chlamydia may be cofactorial for the development of cervical neoplasia [141, 142]. For other infections such as syphilis, gonorrhea, bacterial vaginosis and infectious agents such as cytomegalovirus and Epstein-Barr virus no decisive data are available [143].

Age

In a hospital population of 4,144 women with negative cytology, HPV prevalence was measured with general primer PCR and showed the highest HPV positivity rate at a young age with a maximum between 20 and 24 years and a steady decline with increasing age (fig. 5) [144]. The age distribution is similar for all HPV types which is indirect evidence that a cohort effect – lower HPV infection rates in the past and, thus, later in the elderly – may not be responsible for this phenomenon. The HPV-age association also remains significant when adjusted for other risk factors [127]. Age dependency of HPV detection may be explained by acquisition of immunity or hormonal changes with increasing age. Since even with the highly sensitive PCR, the prevalence of high-risk HPV is very low in patients older than 40 years, this age group may be a suitable target group for HPV screening programs since the specificity of HPV for the detection of SILs would be higher in this cohort [C. Meijer, pers. commun.]. Age and HPV positivity are also inversely correlated in cancer patients: of patients with squamous cell carcinoma (n = 171), 82.5% were HPV positive by in situ hybridization showing a mean age of 51.9 years; of the 17.5% HPV-negative patients, the mean age was 63.5 years [145]. For patients with adeno- or adenosquamous carcinoma (n = 41), a significant difference was again seen for the mean ages of HPV positives (70.7%) at 44.4 years and HPV negatives (29.3%) at 60.1 years. In vulvar cancer, the mean ages of patients with well-differentiated keratinizing squamous cell carcinoma varied between 73 and 77 years with 10% of tumors being HPV positive in contrast to the intraepithelial or basaloid-like squamous cell carcinoma where the mean ages of patients were between 47 and 64 years and 80% were positive for HPV [146, 147].

Pregnancy and the Menstrual Cycle

An increased prevalence of HPV infection has been reported in several studies of pregnant compared to nonpregnant women [148–150; K.H. Fife, pers. commun.]. This observation was not confirmed by others [151–153; J. Chang-Caude, pers. commun.] and controlling for age neutralized an association observed between 624 pregnant women and 2,948 controls using general primer PCR for HPV detection [J.M.M. Walboomers, pers. commun.]. Significant differences in the prevalence of high-risk HPV types but not of low- risk HPV types in pregnant compared with nonpregnant women were reported recently [K.H. Fife, pers. commun.]. An increase of HPV infections during pregnancy followed by a decrease postpartum is also the subject of controversy [148, 149, 153, 154]. Early age at first delivery seems to be an independent risk factor for the development of cervical cancer (OR: 5.0) when comparing women younger than 16 years with women older than 24 years in a consensus PCR-based case-control study [128]. Association of parity with cervical cancer [123] may be related to the development of HPV-associated neoplasia and not infection with HPV since adjustment for HPV-detection increased relative risks observed in a recent case-control study (fig. 4) [129]. Thus, exposure of the transformation zone to a high HPV load at the end of pregnancy and to repeated traumata during multiple births at a young age, when immunity against HPV is immature, may be required for induction of cervical neoplasia.

With respect to the menstrual cycle, a significantly higher prevalence of HPV-16-positive smears was found in the luteal phase compared with the follicular phase in a longitudinal study of 21 women by PCR [68]. No association with the stage of the menstrual cycle was found in cross-sectional studies of 427 and 1,063 nonpregnant women analyzed by filter in situ hybridization [155; Schneider, unpubl. data]. This discrepancy may be explained by differing sensitivity of the HPV test, since even low levels of misclassification of HPV status can distort true relationships between risk factors and HPV [156].

Oral Contraceptives and Hormones

In vitro and in vivo experiments show that hormonal factors may influence transcription and/or translation of the HPV genome. In transgenic mice carrying the *Escherichia coli* β-galactosidase gene under the control of the HPV 16 or HPV 18 LCR, the activity of the reporter gene was shown to be induced after oral administration of dexamethasone but also to be dependent on the physiologic level of steroids in female mice [157; L. Gissmann, pers. commun.]. Treatment of SiHa cells with β-estradiol leads to overexpression of HPV 16 E6 and E7 ORF transcripts [158], and progesterone or progestin treatment transforms primary baby rat kidney cells in the presence of HPV 16 DNA and the *ras* oncogene [159]. In cervical keratinocytes, HPV 16 gene expression can be induced by progesterone and glucocorticoids via three hormone response elements in the regulatory region of the viral genome [160] which may explain the increased expression of progesterone receptors in HPV-16- and HPV-18-positive high-grade SILs [161]. 16α-Hydroxylation of estradiol in keratinocytes derived from normal larynx or laryngeal papillomas increases cell proliferation and can be counteracted by 2-hydroxylation of estradiol to 2-hydroxyestrone, an effect which could be confirmed in HPV-11-infected nude mice treated with indole-3-carbinole, a 2-hydroxylation inducer [162].

In vivo long-term use of oral contraceptives increases the risk for condylomata acuminata [163, 164]. HPV was detected with increasing frequency the longer oral contraceptives were taken [127]. This association was independent of lifetime number of male sexual partners, age, and race. It is not clear if oral contraceptive use is mainly responsible for acquisition of genital HPV infection or transformation of HPV-infected epithelium: an increased risk for cervical cancer with oral contraceptive use was only seen in HPV-positive women [128], whereas in another case-control study, adjustment for HPV neutralized relative risks seen for oral contraceptive use [129].

Smoking

An association between smoking and cervical cancer is postulated by the large majority of more than 30 studies conducted in the past 28 years [165]. However, the biological basis for this association, especially in the context with the strongest risk factor for cervical cancer, i.e. HPV, remains unclear. Cervical mucus of smokers contains elevated concentrations of nicotine and cotinine which may have a transforming effect on HPV-infected tissue [166]. In vitro experiments with HPV-16-immortalized human oral keratinocytes treated by a tobacco-specific N-nitrosamine showed enhanced proliferation capacity in nude mice and increased transcription of HPV 16 E6/E7 genes, epidermal growth factor (EGF) receptors, and the c-*myc* gene when compared with the immortalized counterpart cell line [167]. In humans, the number of Langerhans' cells decreases in cervical tissues of smokers [168] which may promote acquisition of HPV infection. Individuals with a history of smoking have an increased risk for condylomata acuminata [164]. A dose-dependent association between cigarette smoking and the presence of high-risk HPV types as detected by consensus PCR in cervical smears of 181 women remained significant after adjustment for age at first intercourse and lifetime number of sexual partners [169]. That smoking acts synergistically with HPV for cervical carcinogenesis or just increases the risk for HPV infection awaits further clarification: in a case-control study, adjustment for HPV could decrease relative risks between smoking and SILs [129], whereas in another study an association was only found in HPV-positive women [170].

Nutritional Factors

HPV-associated intraepithelial and invasive neoplasia are associated with deficiency of antioxidants and folic acid [171]. Low folate level was an independent risk factor for HPV 16 infection as detected by Southern blot hybridization [172]. However, the therapeutic influence of nutritional factors on HPV infections has not been proven [173], although in vitro and animal studies show that all-trans-retinoic acid treatment may lead to bovine papillomavirus elimination, reversal of transformation in mouse cells [174], and regression of CRPV-infected papillomas [175]. Retinoid acid treatment inhibits HPV 18 E6 and E7 expression in HeLa cells and a lack of retinoic acid receptor (RAR)-β gene expression in tumorigenic hybrid segregants is evidence that insufficient RAR-β expression may contribute to cervical cancer development [176]. In addition, retinoid acid treatment of HPV-16-immortalized ectocervical cells reduces EGF binding and EGF receptor protein levels which is not seen in normal or HPV-negative cervical cancer cell lines [177].

Genetic Predisposition

It is unclear if certain HLA types are associated with higher risk for HPV-associated neoplasia: presence of the HLA class II DQw3 antigen was associated with 7-fold increased risk for cervical cancer, and the presence of the DR6 antigen showed a protective effect [178]. This was not confirmed by others [89].

The presented results on studies of risk factors differ in their validity since adjustment for the different risk factors of HPV infection or HPV-associated neoplasia is incomplete. For future studies it seems important to focus on factors which are responsible for the persistence of HPV infection and continuous expression of viral proteins such as E6 and E7.

References

1 Muñoz N, Bosch FX: HPV and cervical neoplasia: Review of case-control and cohort studies; in Muñoz N, Bosch FX, Shah KV, Meheus A (eds): The Epidemiology of Cervical Cancer and Human Papillomavirus. Lyon, IARC, 1992, pp 251–261.

2 Barr BB, Benton EC, McLaren K, Bunney MH, Smith IW, Blessing K, Hunter JA: Human papilloma virus infection and skin cancer in renal allograft recipients. Lancet 1989;i:124–129.

3 Deau MC, Favre M, Orth G: Genetic heterogeneity among human papillomaviruses (HPV) associated with epidermodysplasia verruciformis: Evidence for multiple allelic forms of HPV5 and HPV8 E6 genes. Virology 1991;184:492–503.

4 Schneider A, Koutsky LA: Natural and epidemiological features of genital HPV infection; in Muñoz N, Bosch FX, Shah KV, Meheus A (eds): The Epidemiology of Cervical Cancer and Human Papillomavirus. Lyon, IARC, 1992, pp 25–52.

5 Morrison EA, Burk RD: Classification of human papillomavirus infection (letter). JAMA 1993;270:453.

6 Crum CP, Ikenberg H, Richart RM, Gissmann L: Human papillomvirus type 16 and early cervical neoplasia. N Engl J Med 1984;310:880–883.

7 Robertson AJ, Anderson JM, Beck JS, et al: Observer variability in histopathological reporting of cervical biopsy specimens. J Clin Pathol 1989;42:231–238.

8 Tierney RJ, Ellis JR, Winter H, Kaur S, Wilson S, Woodman CB, Young LS: PCR for the detection of cervical HPV16 infection: The need for standardization (letter). Int J Cancer 1993;54:700–701.

9 Schiffman MH, Bauer HM, Lorincz AT, Manos MM, Byrne JC, Glass AG, Cadell DM, Howley PM: Comparison of Southern blot hybridization and polymerase chain reaction methods for the detection of human papillomavirus DNA. J Clin Microbiol 1991;29:573–577.

10 Schiffman MH: Validation of hybridization assays: Correlation of filter in situ, dot blot and PCR with Southern blot; in Muñoz N, Bosch FX, Shah KV, Meheus A (eds): The Epidemiology of Cervical Cancer and Human Papillomavirus. Lyon, IARC, 1992, pp 169–179.

11 Lorincz A: Detection of human papillomavirus DNA without amplification: Prospects for clinical utility; in Muñoz N, Bosch FX, Shah KV, Meheus A (eds): The Epidemiology of Cervical Cancer and Human Papillomavirus. Lyon, IARC, 1992, pp 135–145.
12 Gravitt PE, Manos MM: Polymerase chain reaction-based methods for the detection of human papillomavirus DNA; in Muñoz N, Bosch FX, Shah KV, Meheus A (eds): The Epidemiology of Cervical Cancer and Human Papillomavirus. Lyon, IARC, 1992, pp 121–133.
13 Snijders PJ, van den Brule AJ, Schrijnemakers HF, Snow G, Meijer CJ, Walboomers JM: The use of general primers in the polymerase chain reaction permits the detection of a broad spectrum of human papillomavirus genotypes. J Gen Virol 1990;71:173–181.
14 Manos MM, Ting Y, Wright DK, Lewis AJ, Broker TR, Wolinsky SM: The use of polymerase chain reaction amplification for the detection of genital human papillomaviruses. Cancer Cells 1989;7:209–214.
15 Bergeron C, Barrasso R, Beaudenon S, Flamant P, Croissant O, Orth G: Human papillomaviruses associated with cervical intraepithelial neoplasia: Great diversity and distinct distribution in low- and high-grade lesions. Am J Surg Pathol 1992;16:641–649.
16 Lungu O, Sun XW, Felix J, Richart RM, Silverstein S, Wright T Jr: Relationship of human papillomavirus type to grade of cervical intraepithelial neoplasia. JAMA 1992;267:2493–2496.
17 van den Brule AJ, Walboomers JM, Du-Maine M, Kenemans P, Meijer C: Difference in prevalence of human papillomavirus genotypes in cytomorphologically normal cervical smears is associated with a history of cervical intraepithelial neoplasia. Int J Cancer 1991;48:404–408.
18 Resnick RM, Cornelissen MT, Wright DK, Eichinger GH, Fox HS, ter-Schegget J, Manos MM: Detection and typing of human papillomavirus in archival cervical cancer specimens by DNA amplification with consensus primers. J Natl Cancer Inst 1990;82:1477–1484.
19 Das BC, Sharma JK, Gopalkrishna V, Das DK, Singh V, Gissmann L, zur-Hausen H, Luthra UK: A high frequency of human papillomavirus DNA sequences in cervical carcinomas of Indian women as revealed by Southern blot hybridization and polymerase chain reaction. J Med Virol 1992;36:239–245.
20 ter-Meulen J, Eberhardt HC, Luande J, et al: Human papillomavirus (HPV) infection, HIV infection and cervical cancer in Tanzania, East Africa. Int J Cancer 1992;51:515–521.
21 Riou G, Favre M, Jeannel D, Bourhis J, Le-Doussal V, Orth G: Association between poor prognosis in early-stage invasive cervical carcinomas and non-detection of HPV DNA. Lancet 1990;335:1171–1174.
22 Lorincz AT, Reid R, Jenson AB, Greenberg MD, Lancaster W, Kurman RJ: Human papillomavirus infection of the cervix: Relative risk associations of 15 common anogenital types. Obstet Gynecol 1992;79:328–337.
23 Van-Ranst M, Kaplan JB, Burk RD: Phylogenetic classification of human papillomaviruses: Correlation with clinical manifestations. J Gen Virol 1992;73:2653–2660.
24 Li CC, Shah KV, Seth A, Gilden RV: Identification of the human papillomavirus type 6b L1 open reading frame protein in condylomas and corresponding antibodies in human sera. J Virol 1987;61:2684–2690.
25 Jenison SA, Firzlaff JM, Langenberg A, Galloway DA: Identification of immunoreactive antigens of human papillomavirus type 6b by using *Escherichia coli*-expressed fusion proteins. J Virol 1988;62:2115–2123.
26 Strike DG, Bonnez W, Rose RC, Reichman RC: Expression in *Escherichia coli* of seven DNA fragments comprising the complete L1 and L2 open reading frames of human papillomavirus type 6b and localization of the 'common antigen' region. J Gen Virol 1989;70:543–555.
27 Jenison SA, Yu XP, Valentine JM, Koutsky LA, Christiansen AE, Beckmann AM, Galloway DA: Evidence of prevalent genital-type human papillomavirus infections in adults and children. J Infect Dis 1990;162:60–69.
28 Köchel HG, Monazahian M, Sievert K, Hohne M, Thomssen C, Teichmann A, Arendt P, Thomssen R: Occurrence of antibodies to L1, L2, E4 and E7 gene products of human papillomavirus types 6b, 16 and 18 among cervical cancer patients and controls. Int J Cancer 1991;48:682–688.
29 Köchel HG, Sievert K, Monazahian M, Mittelstadt-Deterding A, Teichmann A, Thomssen R: Antibodies to human papillomavirus type-16 in human sera as revealed by the use of prokaryotically expressed viral gene products. Virology 1991;182:644–654.
30 Jochmus-Kudielka I, Schneider A, Braun R, Kimmig R, Koldovsky U, Schneeweis KE, Seedorf K, Gissmann L: Antibodies against the human papillomavirus type 16 early proteins in human sera: Correlation of anti-E7 reactivity with cervical cancer. J Natl Cancer Inst 1989;81:1698–1704.
31 Bleul C, Müller M, Frank R, et al: Human papillomavirus type 18 E6 and E7 antibodies in human sera: Increased anti-E7 prevalence in cervical cancer patients. J Clin Microbiol 1991; 29:1579–1588.
32 Dillner L, Bekassy Z, Jonsson N, Moreno-Lopez J, Blomberg J: Detection of IGA antibodies against human papillomavirus in cervical secretions from patients with cervical intraepithelial neoplasia. Int J Cancer 1989;43: 36–40.
33 Suchankova A, Ritter O, Hirsch I, Krchnak V, Kalos Z, Hamsikova E, Brichacek B, Vonka V: Presence of antibody reactive with synthetic peptide derived from L2 open reading frame of human papillomavirus types 6b and 11 in human sera. Acta Virol (Praha) 1990;34:433–442.
34 Dillner L, Moreno-Lopez J, Dillner J: Serological responses to papillomavirus group-specific antigens in women with neoplasia of the cervix uteri. J Clin Microbiol 1990;28:624–627.
35 Krchnak V, Vagner J, Suchankova A, Krcmar M, Ritterova L, Vonka V: Synthetic peptides derived from E7 region of human papillomavirus type 16 used as antigens in Elisa. J Gen Virol 1990;71:2719–2724.
36 Müller M, Gausepohl H, de-Martynoff G, Frank R, Brasseur R, Gissmann L: Identification of seroreactive regions of the human papillomavirus type 16 protein E4, E6, E7 and L1. J Gen Virol 1990;71:2709–2717.
37 Comerford SA, McCance DJ, Dougan G, Tite JP: Identification of T- and B-cell epitopes of the E7 protein of human papillomavirus type 16. J Virol 1991;65:4681–4690.
38 Selvey LA, Tindle RW, Geysen HM, Haller CJ, Smith JA, Frazer I: Identification of B-epitopes in the human papillomavirus 18 E7 open reading frame protein. J Immunol 1990;145:3105–3110.
39 Tindle RW, Smith JA, Geysen HM, Selvey LA, Frazer IH: Identification of B epitopes in human papillomavirus type 16 E7 open reading frame protein. J Gen Virol 1990;1347–1354.
40 Jenison SA, Yu XP, Valentine JM, Galloway DA: Human antibodies react with an epitope of the human papillomavirus type 6b L1 open reading frame which is distinct from the type-common epitope. J Virol 1989;63:809–818.
41 Mann VM, de-Lao SL, Brenes M, Brinton LA, Rawls JA, Green M, Reeves WC, Rawls WE: Occurrence of IGA and IGG antibodies to select peptides representing human papillomavirus type 16 among cervical cancer cases and controls. Cancer Res 1990;50:7815–7819.
42 Reeves WC, Rawls JA, Green M, Rawls WE: Antibodies to human papillomavirus type 16 in patients with cervical neoplasia (letter). Lancet 1990;335:551–552.
43 Viscidi RP, Sun Y, Tsuzaki B, Bosch FX, Muñoz N, Shah KV: Serologic response in human papillomavirus-associated invasive cervical cancer. Cancer 1993;55:780–784.
44 Müller M, Viscidi RP, Sun Y, Guerrero E, Hill PM, Shah F, Bosch FX, Muñoz N, Gissmann L, Shah KV: Antibodies to HPV-16 E6 and E7 proteins as markers for HPV-16-associated invasive cervical cancer. Virology 1992;187:508–514.
45 Steele JC, Gallimore PH: Humoral assays of human sera to disrupted and nondisrupted epitopes of human papillomavirus type 1. Virology 1990;174:388–398.
46 Bonnez W, Da-Rin C, Rose RC, Reichman RC: Use of human papillomavirus type 11 virions in an Elisa to detect specific antibodies in humans with condylomata acuminata. J Gen Virol 1991;72:1343–1347.
47 Bonnez W, Kashima HK, Leventhal B, Mounts P, Rose RC, Reichman RC, Shah KV: Antibody response to human papillomavirus (HPV) type 11 in children with juvenile-onset recurrent respiratory papillomatosis (RRP). Virology 1992;188:384–387.
48 Kirnbauer R, Booy F, Cheng N, Lowy DR, Schiller JT: Papillomavirus L1 major capsid protein self-assembles into virus-like particles that are highly immunogenic. Proc Natl Acad Sci USA 1992;89:12180–12184.

49 Lowhagen GB, Bolmstedt A, Ryd W, Voog E: The prevalence of 'high-risk' HPV types in penile condyloma-like lesions: Correlation between HPV type and morphology. Genitourin Med 1993;69:87–90.
50 Demeter LM, Stoler MH, Bonnez W, Corey L, Pappas P, Strussenberg J, Reichman RC: Penile intraepithelial neoplasia: Clinical presentation and an analysis of the physical state of human papillomavirus DNA. J Infect Dis 1993;168:38–46.
51 Jablonska S, Orth G, Obalek S, Croissant O: Cutaneous warts: Clinical, histologic, and virologic correlations. Clin Dermatol 1985;3:71–82.
52 Crum CP, Mitao M, Levine RU, Silverstein S: Cervical papillomaviruses segregate within morphologically distinct precancerous lesions. J Virol 1985;54:675–681.
53 Genest DR, Stein L, Cibas E, Sheets E, Zitz JC, Crum CP: A binary (Bethesda) system for classifying cervical cancer precursors: Criteria, reproducibility, and viral correlates. Hum Pathol 1993;24:730–736.
54 Mazzatenta C, Andreassi L, Biagioli M, Ricci S, Ratti G: Detection and typing of genital papillomaviruses in men with a single polymerase chain reaction and type-specific DNA probes. J Am Acad Dermatol 1993;28:704–710.
55 Schneider A, Meinhardt G, Kirchmayr R, Schneider V: Prevalence of human papillomavirus genomes in tissues from the lower genital tract as detected by molecular in situ hybridization. Int J Gynecol Pathol 1991;10:1–14.
56 Nuovo GJ, Friedman D, Richart RM: In situ hybridization analysis of human papillomavirus DNA segregation patterns in lesions of the female genital tract. Gynecol Oncol 1990;36: 256–262.
57 Nuovo GJ, Blanco JS, Leipzig S, Smith D: Human papillomavirus detection in cervical lesions nondiagnostic for cervical intraepithelial neoplasia: Correlation with Papanicolaou smear, colposcopy, and occurrence of cervical intraepithelial neoplasia. Obstet Gynecol 1990; 75:1006–1011.
58 Stoler MH, Whitbeck A, Wolinsky SM, Broker TR, Chow LT, Howett MK, Kreider JW: Infectious cycle of human papillomavirus type 11 in human foreskin xenografts in nude mice. J Virol 1990;64:3310–3318.
59 Stoler MH, Whitbeck A: Differentiation-linked human papillomavirus types 6 and 11 transcription in genital condylomata revealed by in situ hybridization with message-specific RNA probes. Virology 1989;172:331–340.
60 Nuovo GJ, Hochman HA, Eliezri YD, Lastarria D, Comite SL, Silvers DN: Detection of human papillomavirus DNA in penile lesions histologically negative for condylomata: Analysis by in situ hybridization and the polymerase chain reaction. Am J Surg Pathol 1990;14:829–836.
61 Scott GM, Csonka GW: Effect of injections of small doses of human fibroblast interferon into genital warts: A pilot study. Br J Vener Dis 1979;55:442–445.
62 Schonfeld A, Nitke S, Schattner A, et al: Intramuscular human interferon-beta injections in treatment of condylomata acuminata. Lancet 1984;i:1038–1042.
63 Eron LJ, Judson F, Tucker S, et al: Interferon therapy for condylomata acuminata. New Engl J Med 1986;315:1059–1064.
64 Keay S, Teng N, Eisenberg M, Story B, Sellers PW, Merigan TC: Topical interferon for treating condyloma acuminata in women. J Infect Dis 1988;158:934–939.
65 Ferenczy A, Mitao M, Nagai N, Silverstein SJ, Crum CP: Latent papillomavirus and recurring genital warts. N Engl J Med 1985;313:784–788.
66 Pignatari S, Smith EM, Gray SD, Shive C, Turek LP: Detection of human papillomavirus infection in diseased and nondiseased sites of the respiratory tract in recurrent respiratory papillomatosis patients by DNA hybridization. Ann Otol Rhinol Laryngol 1992;101:408–412.
67 Rosenfeld WD, Rose E, Vermund SH, Schreiber K, Burk RD: Follow-up evaluation of cervicovaginal human papillomavirus infection in adolescents. J Pediatr 1992;121:307–311.
68 Schneider A, Kirchhoff T, Meinhardt G, Gissmann L: Repeated evaluation of human papillomavirus 16 status in cervical swabs of young women with a history of normal Papanicolaou smears. Obstet Gynecol 1992;79:683–688.
69 Moscicki AB, Palefsky J, Smith G, Siboshski S, Schoolnik G: Variability of human papillomavirus DNA testing in a longitudinal cohort of young women. Obstet Gynecol 1993;82:578–585.
70 Koutsky LA, Holmes KK, Critchlow CW, Stevens CE, Paavonen J, Beckmann AM, DeRouen TA, Galloway DA, Vernon D, Kiviat NB: A cohort study of the risk of cervical intraepithelial neoplasia grade 2 or 3 in relation to papillomavirus infection. N Engl J Med 1992;327: 1272–1278.
71 Falcinelli C, Claas E, Kleter B, Quint WG: Detection of the human papilloma virus type 16 mRNA-transcripts in cytological abnormal scrapings. J Med Virol 1992;37:93–98.
72 Hsu EM, McNicol PJ, Guijon FB, Paraskevas M: Quantification of HPV-16 E6-E7 transcription in cervical intraepithelial neoplasia by reverse transcriptase polymerase chain reaction. Int J Cancer 1993;55:397–401.
73 Eschle D, Dürst M, ter-Meulen J, Luande J, Eberhardt HC, Pawlita M, Gissmann L: Geographical dependence of sequence variation in the E7 gene of human papillomavirus type 16. J Gen Virol 1992;73:1829–1832.
74 Kiviat NB, Koutsky LA: Specific human papillomavirus types as the causal agents of most cervical intraepithelial neoplasia: Implications for current views and treatment (editorial). J Natl Cancer Inst 1993;85:934–935.
75 Carson HJ, DeMay RM: The mode ages of women with cervical dysplasia. Obstet Gynecol 1993;82:430–434.
76 Aiba S, Rokugo M, Tagami H: Immunohistologic analysis of the phenomenon of spontaneous regression of numerous flat warts. Cancer 1986;58:1246–1251.
77 Rogozinski TT, Jablonska S, Jarzabek-Chorzelska M: Role of cell-mediated immunity in spontaneous regression of plane warts. Int J Dermatol 1988;27:322–326.
78 Fierlbeck G, Rassner G: Condylomata acuminata gigantea mit Nachweis von HPV-6-DNA: Fallbericht mit adjuvanter systemischer IFN-gamma-Therapie. Hautarzt 1989;40:767–770.
79 Hoepfl RM, Christensen ND, Angell MG, Kreider JW: Skin test to assess immunity against cottontail rabbit papillomavirus antigens in rabbits with progressing papillomas or after papilloma regression. J Invest Dermatol 1993;101:227–231.
80 Carson LF, Twiggs LB, Fukushima M, Ostrow RS, Faras AJ, Okagaki T: Human genital papilloma infections: An evaluation of immunologic competence in the genital neoplasia-papilloma syndrome. Am J Obstet Gynecol 1986;155: 784–789.
81 Cauda R, Tyring SK, Grossi CE, Tilden AB, Hatch KD, Sams WM Jr, Baron S, Whitley RJ: Patients with condyloma acuminatum exhibit decreased interleukin-2 and interferon gamma production and depressed natural killer activity. J Clin Immunol 1987;7:304–311.
82 Tay SK, Jenkins D, Maddox P, Campion M, Singer A: Subpopulations of Langerhans' cells in cervical neoplasia. Br J Obstet Gynaecol 1987;94:10–15.
83 Hughes RG, Norval M, Howie SE: Expression of major histocompatibility class II antigens by Langerhans' cells in cervical intraepithelial neoplasia. J Clin Pathol 1988;41:253–259.
84 Hawthorn RJ, Murdoch JB, MacLean AB, MacKie RM: Langerhans' cells and subtypes of human papillomavirus in cervical intraepithelial neoplasia. BMJ 1988;297:643–646.
85 Viac J, Guerin-Reverchon I, Chardonnet Y, Bremond A: Langerhans' cells and epithelial cell modifications in cervical intraepithelial neoplasia: Correlation with human papillomavirus infection. Immunobiology 1990;180: 328–338.
86 Malejczyk J, Majewski S, Jablonska S, Rogozinski TT, Orth G: Abrogated NK-cell lysis of human papillomavirus (HPV)-16-bearing keratinocytes in patients with pre-cancerous and cancerous HPV-induced anogenital lesions. Int J Cancer 1989;43:209–214.
87 Malejczyk J, Malejczyk M, Majewski S, Orth G, Jablonska S: NK-cell activity in patients with HPV16-associated anogenital tumors: Defective recognition of HPV16-harboring keratinocytes and restricted unresponsiveness to immunostimulatory cytokines. Int J Cancer 1993; 54:917–921.
88 Evans CA, Thomsen JJ: Antitumor immunity in the Shope papilloma-carcinoma complex. IV: Search for a transmissible factor increasing the frequency of tumor regression. J Natl Cancer Inst 1993;42:477–484.
89 Glew SS, Duggan-Keen M, Cabrera T, Stern PL: HLA class II antigen expression in human papillomavirus-associated cervical cancer. Cancer Res 1992;52:4009–4016.

90 Warhol MJ, Gee B: The expression of histocompatibility antigen HLA-DR in cervical squamous epithelium infected with human papillomavirus. Mod Pathol 1989;2:101–104.
91 Cromme FV, Meijer CJ, Snijders PJ, Uyterlinde A, Kenemans P, Helmerhorst T, Stern PL, van-den Brule AJ, Walboomers JM: Analysis of MHC class I and II expression in relation to presence of HPV genotypes in premalignant and malignant cervical lesions. Br J Cancer 1993;67:1372–1380.
92 Glew SS, Connor ME, Snijders PJ, Stanbridge CM, Buckley CH, Walboomers JM, Meijer CJ, Stern PL: HLA expression in preinvasive cervical neoplasia in relation to human papilloma virus infecetion. Eur J Cancer 1993;29:1963–1970.
93 Höpfl R, Sandbichler M, Sepp N, et al: Skin test for HPV type 16 proteins in cervical intraepithelial neoplasia (letter). Lancet 1991; 337:373–374.
94 Bernstein A, Harris B: The relationship of dietary and serum vitamin A to the occurrence of cervical intraepithelial neoplasia in sexually active women. Am J Obstet Gynecol 1984;148:309–312.
95 Jarrett WF, Smith KT, O'Neil BW, Gaukroger JM, Chandrachud L, Grindlay GJ, McGarvie GM, Campo MS: Studies on vaccination against papillomaviruses. Prophylactic and therapeutic vaccination with recombinant structural proteins. Virology 1991;184:33–42.
96 Lin YL, Borenstein LA, Selvakumar R, Ahmed R, Wettstein FO: Effective vaccination against papilloma development by immunization with L1 or L2 structural protein of cottontail rabbit papillomavirus. Virology 1992;187:612–619.
97 Lin YL, Borenstein LA, Ahmed R, Wettstein FO: Cottontail rabbit papillomavirus L1 protein-based vaccines: Protection is achieved only with a full-length, nondenatured product. J Virol 1993;67:4154–4162.
98 Barrett TJ, Silbar JD, McGinley JP: Genital warts – A venereal disease. JAMA 1954;154: 333–334.
99 Oriel JD: Natural history of genital warts. Br J Vener Dis 1971;47:1–13.
100 Barrasso R, de Brux J, Croissant O, Orth G: High prevalence of papillomavirus-associated penile intraepithelial neoplasia in sexual partners of women with cervical intraepithelial neoplasia. N Engl J Med 1987;317:916–923.
101 Schneider A, Kirchmayr R, De Villiers EM, Gissmann L: Subclinical human papillomavirus infections in male sexual partners of female carriers. J Urol 1988;140:1431–1434.
102 Ho L, Tay SK, Chan SY, Bernard HU: Sequence variants of human papillomavirus type 16 from couples suggest sexual transmission with low infectivity and polyclonality in genital neoplasia. J Infect Dis 1993;168:803–809.
103 Green J, Monteiro E, Bolton VN, Sanders P, Gibson PE: Detection of human papillomavirus DNA by PCR in semen from patients with and without penile warts. Genitourin Med 1991;67:207–210.
104 Nieminen P, Koskimies AI, Paavonen J: Human papillomavirus DNA is not transmitted by semen. Int J STD AIDS 1991;2:207–208.
105 van-Doornum GJ, Hooykaas C, Juffermans LH, van der Lans SM, van der Linden MM, Coutinho RA, Quint WG: Prevalence of human papillomavirus infections among heterosexual men and women with multiple sexual partners. J Med Virol 1992;37:13–21.
106 Della-Torre G, Donghi R, de Campos Lima PO, Pasquini G, Pilotti S, Koronel R, Pierotti MA, De Palo G, Della-Porta G, Rilke F: Human papillomavirus genomes in male urethral cells. Am J Pathol 1992;141:1181–1186.
107 Hillman RJ, Ryait BK, Botcherby M, Walker MM, Taylor-Robinson D: Human papillomavirus DNA in the urogenital tracts of men with genital dermatoses: Evidence for multifocal infection. Int J STD AIDS 1993;4:147–154.
108 Forslund O, Hansson BG, Rymark P, Bjerre B: Human papillomavirus DNA in urine samples compared with that in simultaneously collected urethra and cervix samples. J Clin Microbiol 1993;31:1975–1979.
109 Geddy PM, Wels M, Lacey CJN: Lack of detection of human papillomavirus DNA in male urine samples. Genitourin Med 1993; 69:276–279.
110 Furihata M, Inoue K, Ohtsuki Y, Hashimoto H, Terao N, Fujita Y: High-risk human papillomavirus infections and overexpression of p53 protein as prognostic indicators in transitional cell carcinoma of the urinary bladder. Cancer Res 1993;53:4823–4827.
111 Kreider JW, Howett MK, Stoler MH, Zaino RJ, Welsh P: Susceptibility of various human tissues to transformation in vivo with human papillomavirus type 11. Int J Cancer 1987;39: 459–465.
112 Shah K, Kashima H, Polk BF, Shah F, Abbey H, Abramson A: Rarity of cesarean delivery in cases of juvenile-onset respiratory papillomatosis. Obstet Gynecol 1986;68:795–799.
113 Sedlacek TV, Lindheim S, Eder C, Hasty L, Woodland M, Ludomirsky A, Rando RF: Mechanism for human papillomavirus transmission at birth. Am J Obstet Gynecol 1989; 161:55–59.
114 Fife KH, Bubalo F, Boggs DL, Gaebler JW: Perinatal exposure of newborns to HPV: Detection by DNA amplification. UCLA Symp Mol Cell Biol 1990;124:73–76.
115 Smith EM, Johnson SR, Cripe TP, Pignatari S, Turek L: Perinatal vertical transmission of human papillomavirus and subsequent development of respiratory tract papillomatosis. Ann Otol Rhinol Laryngol 1991;100:479–483.
116 Chen S, Slavin J, Fairley CK, Tabrizi SN, Borg AJ, Billson V, Garland SM: The absence of HPV DNA in genital specimens from infants. Genitourin Med 1993;69:270–272.
117 Snijders PJ, van den Brule AJ: Schrijnemakers HF, Raaphorst PM, Meijer CJ, Walboomers JM: Human papillomavirus type 33 in a tonsillar carcinoma generates its putative E7 mRNA via two E6* transcript species which are terminated at different early region poly(A) sites. J Virol 1992;66:3172–3178.
118 Chang F, Syrjaenen S, Shen Q, Wang L, Syrjaenen K: Screening for human papillomavirus infections in esophageal squamous cell carcinomas by in situ hybridization. Cancer 1993;72:2525–2530.
119 Moy RL, Eliezri YD, Nuovo GJ, Zitelli JA, Bennett RG, Silverstein S: Human papillomavirus type 16 DNA in periungual squamous cell carcinomas. JAMA 1989;261: 2669–2673.
120 Ruedlinger R, Grob R, Yu YX, Schnyder UW: Human papillomavirus-35-positive bowenoid papulosis of the anogenital area and concurrent human papillomavirus-35-positive verruca with bowenoid dysplasia of the periungual area. Arch Dermatol 1989; 125:655–659.
121 Ferenczy A, Bergeron C, Richart RM: Human papillomavirus DNA in fomites on objects used for the management of patients with genital human papillomavirus infections. Obstet Gynecol 1989;74:950–954.
122 Schneider A, De Villiers EM, Schneider V: Multifocal squamous neoplasia of the female genital tract: Significance of human papillomavirus infection of the vagina after hysterectomy. Obstet Gynecol 1987;70:294–298.
123 Brinton LA: Epidemiology of cervical cancer – Overview; in Muñoz N, Bosch FX, Shah KV, Meheus A (eds): The Epidemiology of Cervical Cancer and Human Papillomaviruses. Lyon, IARC, 1992, pp 3–23.
124 Rosenfeld WDV, Wentz SJ: High prevalence rate of human papillomavirus infection and association with abnormal Papanicolaou smears in sexually active adolescents. Am J Dis Child 1989;143:1443–1447.
125 Moscicki AB, Palefsky J, Gonzales J, Schoolnik GK: Human papillomavirus infection in sexually active adolescent females: Prevalence and risk factors. Pediatr Res 1990;28: 507–513.
126 Fisher M, Rosenfeld WD, Burk RD: Cervicovaginal human papillomavirus infection in suburban adolescents and young adults. J Pediatr 1991;119:821–825.
127 Ley C, Bauer HM, Reingold A, Schiffman MH, Chambers JC, Tashiro CJ, Manos MM: Determinants of genital human papillomavirus infection in young women. J Natl Cancer Inst 1991;83:997–1003.
128 Bosch FX, Munõz N, de Sanjose S, et al: Risk factors for cervical cancer in Colombia and Spain. Int J Cancer 1992;52:750–758.
129 Schiffman MH, Bauer HM, Hoover RN, et al: Epidemiologic evidence showing that human papillomavirus infection causes most cervical intraepithelial neoplasia. J Natl Cancer Inst 1993;85:958–964.

130 Schneider V, Kay S, Lee HM: Immunosuppression as a high-risk factor in the development of condyloma acuminatum and squamous neoplasia of the cervix. Acta Cytol 1983;27:220–224.
131 Halpert R, Fruchter RG, Sedlis A, Butt K, Boyce JG, Sillman FH: Human papillomvirus and lower genital neoplasia in renal transplant patients. Obstet Gynecol 1986;68:251–258.
132 Alloub MI, Barr BB, McLaren KM, Smith IW, Bunney MH, Smart GE: Human papillomavirus infection and cervical intraepithelial neoplasia in women with renal allografts. BMJ 1989;298:153–156.
133 Lewensohn-Fuchs I, Wester D, Bistoletti P, Elfgren K, Ohlman S, Dillner J, Dalianis T: Serological responses to human papillomavirus type 16 antigens in women before and after renal transplantation. J Med Virol 1993; 40:188–192.
134 Johnson JC, Burnett AF, Willet GD, Young MA, Doniger J: High frequency of latent and clinical human papillomavirus cervical infections in immunocompromised human immunodeficiency virus-infected women. Obstet Gynecol 1992;79:321–327.
135 Vermund SH, Kelley KF, Klein RS, Feingold AR, Schreiber K, Munk G, Burk RD: High risk of human papillomavirus infection and cervical squamous intraepithelial lesions among women with symptomatic human immunodeficiency virus infection. Am J Obstet Gynecol 1991;165:392–400.
136 Van Doornum GJ, Van den Hoek JA, Van Ameijden EJ, Van Haastrecht HJ, Roos MT, Henquet CJ, Quint WG, Coutinho RA: Cervical HPV infection among HIV-infected prostitutes addicted to hard drugs. J Med Virol 1993;41:185–190.
137 Caussy D, Goedert JJ, Palefsky J, et al: Interaction of human immunodeficiency and papilloma viruses: Association with anal epithelial abnormality in homosexual men. Int J Cancer 1990;46:214–219.
138 Critchlow CW, Holmes KK, Wood R, Krueger L, Dunphy C, Vernon D, Daling JR, Kiviat NB: Association of human immunodeficiency virus and anal human papillomavirus infection among homosexual men. Arch Intern Med 1992;152:1673–1676.
139 Melbye M, Palefsky J, Gonzales J, Ryder LP, Nielsen H, Bergmann O, Pindborg J, Biggar RJ: Immune status as a determinant of human papillomavirus detection and its association with anal epithelial abnormalities. Int J Cancer 1990;46:203–206.
140 Tornesello ML, Buonaguro FM, Beth-Giraldo E, Giraldo G: Human immunodeficiency virus type 1 *tat* gene enhances human papillomavirus early gene expression. Intervirology 1993;36:57–64.
141 Hildesheim A, Mann V, Brinton LA, Szklo M, Reeves WC, Rawls W: Herpes simplex virus type 2: A possible interaction with human papillomvirus types 16/18 in the development of invasive cervical cancer. Int J Cancer 1991;49:335–340.
142 Koutsky LA, Kiviat NB: Genital infectious agents and invasive cervical cancer (letter). Lancet 1993;342:184–185.
143 Lacey CJ: Assessment of exposure to sexually transmitted agents other than human papillomavirus; in Muñoz N, Bosch FX, Shah KV, Meheus A (eds): The Epidemiology of Cervical Cancer and Human Papillomaviruses. Lyon, IARC, 1992, pp 93–105.
144 Melkert PW, Hopman E, van den Brule AJ, Risse EK, van Diest P, Bleker OP, Helmerhorst T, Schippert ME, Meijer CJ, Walboomers JM: Prevalence of HPV in cytomorphologically normal cervical smears, as determined by the polymerase chain reaction, is age-dependent. Int J Cancer 1993;53:919–923.
145 Higgins GD, Davy M, Roder D, Uzelin DM, Phillips GE, Burrell CJ: Increased age and mortality associated with cervical carcinomas negative for human papillomavirus RNA. Lancet 1991;338:910–913.
146 Andersen WA, Franquemont DW, Williams J, Taylor PT, Crum CP: Vulvar squamous cell carcinoma and papillomaviruses: Two separate entities? Am J Obstet Gynecol 1991;165: 329–335.
147 Toki T, Kurman RJ, Park JS, Kessis T, Daniel RW, Shah KV: Probable nonpapillomavirus etiology of squamous cell carcinoma of the vulva in older women: A clinicopathologic study using in situ hybridization and polymerase chain reaction. Int J Gynecol Pathol 1991;10:107–125.
148 Schneider A, Hotz M, Gissmann L: Increased prevalence of human papillomaviruses in the lower genital tract of pregnant women. Int J Cancer 1987;40:198–201.
149 Czegledy J, Gergely L, Endroedi I: Detection of human papillomavirus deoxyribonucleic acid by filter in situ hybridization during pregnancy. J Med Virol 1989;28:250–254.
150 Villa LL, Franco EL: Epidemiologic correlates of cervical neoplasia and risk of human papillomavirus infection in asymptomatic women in Brazil. J Natl Cancer Inst 1989;81: 332–340.
151 Peng TC, Searle CP, Shah KV, Repke JT, Johnson TR: Prevalence of human papillomavirus infections in term pregnancy. Am J Perinatol 1990;7:189–192.
152 Smith EM, Johnson SR, Jiang D, Zaleski S, Lynch CF, Brundage S, Anderson RD, Turek LP: The association between pregnancy and human papilloma virus prevalence. Cancer Detect Prev 1991;15:397–402.
153 Kemp EA, Hakenewert AM, Laurent SL, Gravitt PE, Stoerker J: Human papillomavirus prevalence in pregnancy. Obstet Gynecol 1992;79:649–656.
154 Rando RF, Lindheim S, Hasty L, Sedlacek TV, Woodland M, Eder C: Increased frequency of detection of human papillomavirus deoxyribonucleic acid in exfoliated cervical cells during pregnancy. Am J Obstet Gynecol 1989;161:50–55.
155 McNicol PJ, Guijon FB, Paraskevas M, Heywood E, Gray MJ, Brunham RC: Effect of the menstrual cycle on detection and typing of human papillomavirus in uterine cervical cells. Am J Obstet Gynecol 1990;162:1037–1041.
156 Franco EL: The sexually transmitted disease model for cervical cancer: Incoherent epidemiologic findings and the role of misclassification of human papillomavirus infection. Epidemiology 1991;2:98–106.
157 Cid A, Auewarakul P, Garcia-Carranca A, Ovseiovich R, Gaissert H, Gissmann L: Cell-type-specific activity of the human papillomavirus type 18 upstream regulatory region in transgenic mice and its modulation by tetradecanoyl phorbol acetate and glucocorticoids. J Virol 1993;67:6742–6752.
158 Mitrani-Rosenbaum ST, Tur Kaspa R: Oestrogen stimulates differential transcription of human papillomavirus type 16 in SiHa cervical carcinoma cells. J Gen Virol 1989;70: 2227–2232.
159 Pater A, Bayatpour M, Pater MM: Oncogenic transformation by human papillomavirus type 16 deoxyribonucleic acid in the presence of progesterone or progestins from oral contraceptives. Am J Obstet Gynecol 1990;162: 1099–1103.
160 Mittal R, Tsutsumi K, Pater A, Pater MM: Human papillomavirus type 16 expression in cervical keratinocytes: Role of progesterone and glucocorticoid hormones. Obstet Gynecol 1993;81:5–12.
161 Monsonego J, Magdalenat H, Catalan F, Coscas Y, Zerat L, Sastre X: Estrogen and progesterone receptors in cervical human papillomavirus related lesions. Int J Cancer 1991;48: 533–539.
162 Newfield L, Goldsmith A, Bradlow HL, Auborn K: Estrogen metabolism and human papillomavirus-induced tumors of the larynx: Chemo-prophylaxis with indole-3-carbinol. Anticancer Res 1993;13:337–341.
163 Franceschi S, Doll R, Gallwey J, La Vecchia C, Peto R, Spriggs AL: Genital warts and cervical neoplasia: An epidemiological study. Br J Cancer 1983;48:621–628.
164 Daling JR, Sherman KJ, Weiss NS: Risk factors for condyloma acuminatum in women. Sex Transm Dis 1986;13:16–18.
165 Winkelstein W Jr: Smoking and cervical cancer – Current status. A review. Am J Epidemiol 1960;131:945–957.
166 Schiffman MH, Haley NJ, Felton JS, Andrews AW, Kaslow RA, Lancaster WD, Kurman RJ, Brinton LA, Lannom LB, Hoffmann D: Biochemical epidemiology of cervical neoplasia: Measuring cigarette smoke constitutents in the cervix. Cancer Res 1997;47: 3886–3888.
167 Kim MS, Shin KH, Baek JH, Cherrick HM, Park NH: HPV-16, tobacco-specific N-nitrosamine, and N-methyl-N′-nitro-N-nitrosoguanidine in oral carcinogenesis. Cancer Res 1993;53:4811–4816.

168 Barton SE, Maddox PH, Jenkins D, Edwards R, Cuzick J, Singer A: Effect of cigarette smoking on cervical epithelial immunity: A mechanism for neoplastic change? Lancet 1988;ii:652–654.
169 Burger MP, Hollema H, Gouw AS, Pieters WJ, Quint WG: Cigarette smoking and human papillomavirus in patients with reported cervical cytological abnormality. BMJ 1993; 306:749–752.
170 Herrero R, Brinton LA, Reeves WC, Brenes MM, Tenorio F, de Britton RC, Gaitan E, Garcia M, Rawis WE: Invasive cervical cancer and smoking in Latin America. J Natl Cancer Inst 1989;81:205–211.
171 Schneider A, Shah K: The role of vitamins in the etiology of cervical neoplasia: An epidemiological review. Arch Gynecol Obstet 1989;246:1–13.
172 Butterworth CE Jr, Hatch KD, Macaluso M, Cole P, Sauberlich HE, Soong SJ, Borst M, Baker VV: Folate deficiency and cervical dysplasia. JAMA 1992;267:528–533.
173 Butterworth CE Jr, Hatch KD, Soong SJ, Cole P, Tamura T, Sauberlich HE, Borst M, Macaluso M, Baker V: Oral folic acid supplementation for cervical dysplasia: A clinical intervention trial. Am J Obstet Gynecol 1992;166: 803–809.
174 Tsang SS, Li G, Stich HF: Effect of retinoic acid on bovine papillomavirus (BPV) DNA-induced transformation and number of BPV DNA copies. Int J Cancer 1988;42:94–98.
175 McMichael H: Inhibition of growth of Shope rabbit papilloma by hypervitaminosis A. Cancer Res 1965;25:947–955.
176 Bartsch D, Boye B, Baust C, zur Hausen H, Schwarz E: Retinoic acid-mediated repression of human papillomavirus 18 transcription and different ligand regulation of the retinoic acid receptor beta gene in nontumorigenic and tumorigenic Hela hybrid cells. EMBO J 1992;11:2283–2291.
177 Sizemore N, Rorke EA: Human papillomavirus 16 immortalization of normal human ectocervical epithelial cells alters retinoic acid regulation of cell growth and epidermal growth factor receptor expression. Cancer Res 1993;53:4511–4517.
178 Wank R, Schendel DJ, Thomssen C: HLA antigens and cervical carcinoma (letter; comment). Nature 1992;356:22–23.
179 Richart RM: Cervical intraepithelial neoplasia. Pathol Annu 1973;8:301–328.
180 Riotton G, Christofferson WM, Lunt R: Cytology of the Female Genital Tract: International Histological Classification of Tumours. Geneva, WHO, 1973.
181 Anonymous: The 1988 Bethesda system for reporting cervical/vaginal cytological diagnoses: National Cancer Institute Workshop. JAMA 1989;262:931–934.

Intervirology 1994;37:215–225

Susan Gross Fisher

Departments of Obstetrics and Gynecology, and Preventive Medicine and Epidemiology, Loyola University Medical Center, Maywood, Ill., USA

Epidemiology: A Tool for the Study of Human Papillomavirus-Related Carcinogenesis

Key Words
Papillomavirus
Cancer
Cervix
Epidemiology
Anogenital

Introduction

Fueled by the recent advances in molecular technology, the field of human papillomavirus (HPV) research is rapidly expanding. The plethora of new techniques in molecular biology and immunology has provided comprehensive profiles of HPV genome structure and function ans its interaction with the infected host. These sophisticated technologies have yielded persuasive data demonstrating HPV activity in carcinogenesis. Although these laboratory techniques continue to be refined, their development has appreciably enhanced the capabilities of epidemiologic assessment related to risk exposure and disease outcome. Furthermore, these advances have provided the tools to better identify and track the course of disease from its earliest inception. The integration of these specialized techniques into the conduct of well-designed, epidemiologic studies is providing critical data to confirm the etiologic role of HPV in human carcinogenesis.

Inherent in clinical investigations are ethical constraints which justifiably limit the conduct of human experimentation, particularly that which relates to disease cause and/or prevention rather than treatment of existing disease. Without the clear advantage of a randomized clinical trial, such as a vaccination intervention study, to establish the causal role of HPV in cancer, we must rely on traditional epidemiologic investigations. These studies consist of the systematic observation of groups of individuals of interest under unmanipulated, natural conditions. Epidemiologic research offers strong methodologies, including both design and quantitative analysis, applied to human populations without the necessity of extrapolation from laboratory studies or animal models. In contrast, the weakness of traditional epidemiologic research is the necessity to infer rather than prove causality; mechanisms of action are essentially impossible to identify in such investigations [1]. Given this restriction, a claim of causation by most epidemiologists is reserved for the most selective, cohesive, and consistent research evidence.

In this chapter, an overview of the principles of causality as applied to epidemiologic data is presented along with a description of the major study designs implemented in epidemiologic research. Given that the most comprehensive laboratory (experimental) data regarding the etiologic role of HPV in human cancer are available

Susan Gross Fisher, PhD
Departments of Obstetrics and Gynecology,
and Preventive Medicine and Epidemiology
Loyola University Medical Center
Maywood, IL 60153 (USA)

0300–5526/94/0374–0215
$8.00/0

for cervical cancer, a brief summary of the natural history of this cancer is presented and is followed by the epidemiologic data supporting the etiologic role of HPV. In addition, the potential role of HPV in other genital and nongenital cancers is briefly discussed. Particular attention is given to data regarding HPV-related cancers among individuals infected with the human immunodeficiency virus (HIV) and individuals with other immunosuppressive conditions.

Causality

From an epidemiologic perspective, causation is a relationship in which an extrinsic or intrinsic factor plays an essential role in producing disease [2]. Historically, this concept usually reflected a clear and direct relationship between a single factor, often an infectious agent, and a resulting condition or disease. Although the existence of contributory factors, such as the impact of poverty and overcrowding (extrinsic) and malnutrition and alcoholism (intrinsic) on the direct association between the tubercle bacillus and tuberculosis, was recognized, the process of cause and effect was, or at least in retrospect seemed to be, less adulterated. Causation is now perceived as an interdependent process in which multiple factors act in concert to promote disease [3].

The major principles [2, 4] upon which causal inference is based are outlined below.

Strength of the Association

The likelihood of a cause-and-effect relationship increases as does the magnitude of the observed association between the risk factor and disease occurrence. Weak associations often prove to be spurious and explainable by a third variable. Relative risk (RR) is the measurement of choice for reflecting this concept.

Consistency among Investigations

Since it is never possible to achieve the level of control in epidemiologic research that is attainable in the research laboratory, a cause-effect relationship is more likely to exist when a series of studies using different methodologies in a variety of settings has consistent conclusions. Although a different result in a different inquiry certainly cannot be held to refute original evidence, the same results from precisley the same form of inquiry will not greatly strengthen the original evidence [5]. Optimally, diverse studies yielding the same association support the plausibility of a causal interpretation.

Scientific Plausibility

Consistency of the potential association with current knowledge regarding the factor and the disease strongly supports the concept of causation. Lack of a known or postulated mechanism, however, should never hinder consideration regarding the causal nature of a particular association. Such an unexplained association may represent the initial suggestion that ultimately leads to a vast extension of scientific knowledge.

Temporality

For a conclusion of causality to be reasonable, the exposure characteristic or event associated with the disease must precede the occurrence of disease and allow an appropriate time interval for explaining induction and/or latency. Evidence to support this principle necessitates the conduct of prospective, longitudinal studies.

Biologic Gradient

Evidence of increasing risk or severity of disease in the presence of an increasing level of exposure to the proposed exposure factor, i.e. dose response, provides compelling support for a causal association. There are, however, many examples in nature of a threshold response in which response is absent until the stimulus reaches a certain level. In addition, dose response can often be masked by an inability to quantify levels of exposure.

Determination of causality, particularly related to human cancers, is of the utmost importance in understanding the process of tumor initiation, promotion, conversion and progression. Optimal approaches for disease prevention, control and treatment are defined when the etiology of the disease is known. In contrast, incorrect identification of the etiologic factor(s) may lead to fruitless efforts which deflect from effective approaches in disease control.

Epidemiologic Study Designs

In order to identify risk factors that are related to disease occurrence, epidemiologists rely on specific methodologies for the conduct of observational studies. These are categorized as cross-sectional, case-control and cohort studies (table 1).

A cross-sectional study examines the status of a sample of individuals with respect to the presence or absence of a factor (exposure) and an outcome (disease) at the same point in time. Subjects are selected to be representative of the population of interest. Since data are collected simul-

Table 1. Characteristics of epidemiologic study designs

Design	Basis of subject selection	End point variable	End point occurrence before study initiation
Cross-sectional	population of interest	risk factor disease outcome	yes
Case control	disease outcome	risk factor	yes
Prospective cohort	risk factor	disease outcome	no
Historical cohort	risk factor	disease outcome	yes

taneously, the occurrence of disease in relation to exposure cannot be defined, this this design cannot address the principle of temporality. This method is most effective for the establishment of a relationship involving risk factors that cannot be altered/manipulated and do not fluctuate over time, e.g. gender, age, blood type. Generally, this method is optimal for suggesting the presence of an association, and providing support for the conduct of additional investigations. As a preliminary investigation this design can be completed quickly and at minimal cost.

A case-control study is a retrospective design in which subject selection is based on the presence (cases) or absence (controls) of a particular condition or disease. Ideally, the control group is as similar to the cases as possible with the exception of the presence of the disease being studied. Subjects within the two groups are then classified according to the presence or absence of a factor (exposure) of interest. A greater frequency of the factor among the cases than among the controls suggests an increased disease risk with the factor. Similarly, a lower than expected presence of the factor among the cases may suggest a protective effect. This design is most useful when studying a disease which is extremely rare.

A cohort study is the most sophisticated study design in that it is prospective and introduces the least probability of bias. In such a study the subjects are selected/categorized on the basis of the presence or absence of the exposure (risk) factor of interest. All subjects are then followed for a specified period of time, often 10 or more years, to determine the number of subjects who develop the disease of interest. An increased rate of disease among the subjects with the exposure factor supports an association between that risk factor and the development of the disease. This design is particularly useful when the association of interest involves a rare exposure factor. A cohort study optimally addresses temporality, since presence of the risk factor is clearly established before any evidence of disease exists. Although all cohort studies are prospective in design, two approaches may be used. At the initiation of the study, subjects can be categorized as to exposure and followed longitudinally to determine occurrence of disease (prospective cohort), or subjects can be selected, categorized as to exposure, and followed for disease occurrence using purely historical data collection methods (historical cohort). While the former approach mandates a lengthy study, it provides the opportunity for standardized data collection including information regarding potential confounding factors to the exposure-disease association being investigated. The latter approach can be completed in a timely fashion since the disease outcomes have already occurred; however, data are rarely available to address issues related to confounding.

Natural History of Cervical Cancer

Cancer of the uterine cervix is one of the most common malignancies in women worldwide. Of the half million females diagnosed with cervical cancer each year, 50% will ultimately die of this disease. In the US, the incident cases of invasive cancers approach 14,000 annually. Cervical cancer is 6-fold more common in developing countries than in Western societies [6, 7]. Undoubtedly this differential incidence is partly due to differing prevalence of risk factors. While in most Western societies, early detection of disease through cytologic screening using the Pap smear has allowed for the provision of treatment early in the disease process, cervical tumors diagnosed in women from developing countries are frequently at advanced stages of disease and result in a high rate of mortality.

Cancer of the cervix begins as a preinvasive neoplastic alteration in cells [cervical intraepithelial neoplasia (CIN)], which may regress, remain unchanged or progress

Table 2. Types of genital HPV

Low risk:	6, 11, 42, 43, 44
Intermediate risk:	30, 31, 33, 34, 35, 39, 40, 49, 51, 52, 57, 58, 63, 64
High risk:	16, 18, 45, 56

to invasive cancer. Since it is not possible to predict which lesions will progress, all lesions require careful monitoring and/or treatment.

Controlled studies have consistently identified correlates of sexual behavior, namely early onset of sexual activity, multiple sexual partners and promiscuity of the partner as the most critical determinants of cervical cancer risk [8–10]. The disease is essentially nonexistent among celibate females in the US [11]. Cervical cancer most commonly occurs in multiparous women during their reproductive years. It occurs most frequently among individuals of lower socioeconomic status, particularly in blacks and Mexican Americans [12]. Smoking has been linked to cervical cancer; in fact, nicotine derivatives have been detected in the cervical mucus of smokers [13]. The independent significance of smoking, however, remains controversial, in particular due to the correlation of smoking with other sexual behavior factors. Unlike other smoking-related cancers in females, cervical cancer has declined, suggesting that such a relationship is either spurious or veiled by improved patterns of screening and treatment for dysplasia. The use of oral contraceptives has been identified as a potential risk factor for cervical cancer; however, a correlation between the practice of birth control and increased sexual activity may account for this association. The more frequent screening of these individuals during routine physician visits for the purpose of renewing their prescriptions for oral contraceptives may also account for increased cervical cancer detection among these women [14].

Early Evidence of an HPV-Cervical Cancer Association

Although the epidemiologic profile of risk factors for cervical cancer strongly suggested that a sexually transmitted infectious organism may be the most likely etiologic culprit, identification of the responsible organism was elusive for several years. The infectious agent which initially appeared most worthy of consideration was herpes simplex virus (HSV), and an increased prevalence of antibodies to HSV was detected among cervical cancer patients; however, viral DNA sequences were not maintained within the genome of transformed cells in culture [15]. Investigators concluded that although HSV may be capable of initiating a malignant phenotype, its presence was not required for maintenance and propagation of the malignancy.

The first evidence of a link between HPV and cervical cancer came with recognition that the morphologic features of dysplasia were manifestations of HPV infection, previously considered to cause self-limited occurrence of genital warts [16]. The similarities in risk factors for HPV infection and cervical cancer also strongly supported further investigation of this agent. The concept of a causal relationship between HPV and cervical cancer was first introduced by zur Hausen [17]. Using molecular techniques, Gissmann et al. [18] demonstrated a strong association of HPV type 6 with condylomata acuminata and with mild dysplasia. Findings showed, however, that most invasive cervical carcinomas were negative for these HPV types. Eventually, zur Hausen and colleagues went on to clone HPV 16 and 18 directly from cervical carcinomas. They demonstrated the regular presence and the specific expression of HPV 16 and 18 E6 and E7 in the cervical tumor biopsies and identified those DNA sequences as the active viral molecules within the tumor cells [19].

More than 70 types of HPV have now been identified, each having less than 50% homology with any other type. Based on the clinical repercussions of type-specific infections, these viruses have been categorized as low-, moderate-, and high-risk types (table 2). Most epidemiologic studies have focused on HPV 6/11, commonly associated with benign genital warts, and HPV 16/18, strongly correlated with the presence of severe dysplasia and carcinoma.

Epidemiologic Studies of HPV and Cervical Cancer

A wide variety of observations and measurements is required to describe and explain the occurrence of a disease in human populations. Even the most convincing laboratory findings regarding health and disease require the consistent support of clinical studies. In cases in which observational studies are most appropriate, initial inquiries are in the form of cross-sectional and case-control studies. Cohort studies, which tend to be larger, more costly to conduct, and take longer to complete, are done only if the initial studies are favorable or a more optimal design is required to answer the research question.

Before the availability of reliable molecular methods, HPV exposure was assessed using inaccurate methods which most commonly consisted of a self-reported history of genital warts and the presence of morphologic signs of HPV infection. In addition, variability in collection techniques for cytologic specimens introduced error into the classification of infection. Such crude definitions of exposure and inconsistency in data collection resulted in extremely variable prevalence rates and the generation of conflicting reports regarding HPV association or lack thereof. Although HPV detection is now much more reliable, the rapid introduction of increasingly powerful, i.e. sensitive, methods has allowed for the detection of minute amounts of HPV in human lesions. These lesions are being considered as positive for HPV; however, their behavior and clinical relevance may differ significantly from lesions containing a higher viral load. This enhancement in HPV detection rates has made the acquisition of a cohesive body of comparable findings elusive, slowing a clear understanding of the HPV-cancer association.

Early studies were also hampered by limitations in research design. Relatively small numbers of cases were included in the studies which led to imprecise estimates of risk. Little attention was given to carefully describing the study samples regarding confounding factors such as age and number of sexual partners; interpretation of findings was, therefore, often difficult. Misclassification was a potential problem in that subjects categorized as controls were often not examined to assure that they were free of disease. Only recently have data become available from well-designed case-control studies that have used valid techniques for case definition and exposure classification [20].

Three major findings have been the focus of the majority of early studies: (1) Investigators have consistently detected a high rate of HPV DNA positivity in biopsy specimens of tumors from women diagnosed with cervical cancer. Depending on the technique used for detection, this rate ranged from 42 to as high as 96% [21, 22]. Cytologic abnormalities were only detected in approximately 10–20% of those testing positive for HPV. (2) The prevalence of HPV increases linearly with the severity of cervical dysplasia. (3) Investigators have confirmed an inverse relationship between the presence of HPV 6/11 and level of neoplasia; similarly, the presence of HPV 16/18 is positively associated with severity of dysplasia.

Studies to evaluate different infections such as chlamydia, gonococcus, and genital warts, have demonstrated that only genital warts were significantly more common among women with invasive cervical cancer. Peters et al. [23] reported a greater than 3-fold risk of invasive cancer [odds ratio (OR): 3.7; 95% confidence interval (CI): 1.6–9.0] among women reporting a history of genital warts. This elevation in risk was supported by Slattery et al. [9] among women with invasive cancer or cancer in situ. In a similar study, Brinton et al. [8] detected an increase in risk among women reporting a history of genital warts; however, this elevation in risk was not statistically significant. Jones et al. [10] did not detect such an elevation in risk in their sample of 293 women with cancer in situ (OR: 1.0; 95% CI: 0.4–2.5). In each of these studies, adjustment was made for other major risk factors for cervical cancer. Associations of genital wart history have also been reported in women under the age of 20 diagnosed with CIN; risk was estimated to be as high as 21.1 (95% CI: 5.1–87.2) [24].

In 1991, Kjaer et al. [25] demonstrated that a history of genital warts among the male partners of monogamous women was an important risk factor for cervical cancer in the females. The estimated risk was 17.9 (95% CI: 3.3–98.3). Detection of HPV DNA in the males did not confirm the reported history; however, hybridization assays on male subjects are rarely sensitive enough for accurate viral detection.

Munoz et al. [26] conducted a large case-control study in Columbia and Spain to examine the association between invasive cervical cancer and history of genital warts. Although they demonstrated increases in risk among women in the sample from Spain (OR: 4.9; 95% CI: 1.0–25.6), the risk among those sampled in Columbia was not significantly elevated.

These reports suggest that although genital warts, mostly containing HPV 6, are sometimes associated with cervical cancer, this association is not consistent. Most likely, the presence of these warts represents a surrogate marker of increased sexual activity, which, in turn, provides an increased risk of infection with other HPV types which are oncogenic in nature.

Recognizing the limitations of genital wart history as a means for determining exposure, many recent epidemiologic studies have taken advantage of the new molecular techniques for identifying HPV infection. A particularly ambitious effort by de Villiers et al. [27] involved the review of 9,295 cervical smears from routine screening examinations. Filter in situ hybridization was used to detect HPV DNA. Ten percent of the normal cytologic smears were positive for HPV DNA, whereas 35–40% of those showing CIN I–III were positive for HPV, as were those with invasive cancer. This rate was independent of age. The investigators estimate that the presence of HPV may have been underestimated by a factor of two to three, since filter in situ hybridization is less sensitive than Southern blot techniques.

In a case-control study of 2,627 women studied by Lorincz et al. [28], HPV positivity was detected in 79.3, 23.7 and 6.4% of women with definite cervical disease, borderline atypia and no cytologic abnormality, respectively. A clear association was demonstrated between HPV type and severity of neoplasia, with HPV 16 being most common among high-grade intraepithelial lesions and invasive carcinomas.

Kiviat et al. [29] examined 454 women seen at a sexually transmitted disease (STD) clinic and 545 women attending a college health service. Using dot blot hybridization, HPV was detected in 9.6% (n = 96) of the women. Among the HPV-positive women, 69% had dysplasias compared to a rate of 6% among those testing negative for HPV. Dysplasia was detected in 53% of those testing positive for HPV 6/11 and 41% among those positive for HPV 16/18.

In 1991, Manos et al. [30] conducted a study of 500 women with squamous intraepithelial lesions (SILs) and 500 controls drawn in a nested case-control study from a cohort of 24,000 women. Among those with low-grade SILs, the OR for HPV positivity using the polymerase chain reaction (PCR) was 20 (95% CI: 13.0–26.0) and among those with high-grade SILs the OR was 40 (95% CI: 16.0–105.0).

In 1991, Morrison et al. [31] provided data similar to those of Manos et al. [30]. In a study comparing 85 women with SILs to 70 with normal cytology, high-risk (type 16 and 18) HPV infection was independently predictive of SILs (OR: 17.9; CI: 6.2–51.6) using Southern blotting. In addition, low educational level was independently associated with SILs (OR: 3.4; CI: 1.2–10.1). Morrison et al. [31] also showed an increase in the hybridization signal with increased risk of dysplasia, suggesting a dose-response phenomenon (OR: 1, 16 and 21 with in situ hybridization and 1, 8 and 13 with PCR). These results were supported by Reeves et al. [32] in 1989 in a multicenter study in Latin America of 759 women with invasive cancer compared to 1,467 age-matched controls. They found the prevalence of HPV to be 62% among the cases as compared to 32% in the controls, and demonstrated a linear increase in risk based on the signal intensity from the hybridization technique.

In women from the United Kingdom, Cuzick et al. [33] demonstrated that those with CIN III had a lower risk (OR: 3.2) of genital warts compared to those with CIN II (OR: 7.6) or CIN I (OR: 8.2). These results suggest that HPV 6/11 has an acitve role in the initiation of genital warts and mild dysplasia, but a less important role in the occurrence of severe dysplasias.

HPV type has been associated with tumor progression. HPV 18 appears to be preferentially associated not only with adenocarcinomas but also with more aggressively growing squamous cell cancers of the uterine cervix [34]. Reinforcing the role of HPV in cervical tumors is the finding that HPV 16 DNA has been detected in lymph nodes and pelvic metastases from HPV-16-positive cervical cancer patients [35–37].

These studies provide relatively consistent data supporting an active role for HPV 16/18 in cervical cancer. The role of HPV 6/11 appears to be most related to benign genital warts, but it may serve as a marker for sexual activity as a risk factor for HPV 16/18 infection. Although informative and important for laying a foundation in HPV epidemiology, these cross-sectional and retrospective studies do not address the temporality of exposure and disease.

Cohort Studies

This longitudinal prospective design is ideal for determining the association between HPV and cervical cancer. Type-specific exposure to HPV as well as exposure to other potential risk factors are assessed at study entry in a large group of unselected women who are then followed for occurrence of cervical neoplasia. Problems in awaiting outcome arise, however, since severe dysplasia should ethically be treated before progression of invasive cancer. Surrogate end points such as CIN must then be used or, alternatively, specimens previously collected from women who ultimately develop cancer and matched controls may be tested (historical cohort study). Such data provide convincing evidence related to risk of progression following HPV infection, with minimal introduction of bias.

In one of the earliest studies by Mitchell et al. [38], 846 women with subclinical HPV infection detected by Pap smear were followed prospectively for 6 years. This study revealed a RR for the development of CIN III of 15.6 compared to the general population. Risk was particularly high for those under the age of 25 years. This study is hampered by the lack of an internal control group. Results could be biased by differences between the study group and the general population used a a comparison.

Campion et al. [39] followed a group of 100 women with CIN I for a period of over 2 years. Fifty-six percent of those with high-risk HPV progressed to CIN III compared with 20% of those with HPV type 6. Similar results were reported by Schneider et al. [40], with a 21% rate of progression over a 6-month follow-up period among CIN I women infected with HPV 16/18 compared to 0% progression among those testing positive for only HPV 6/11.

In a nested case-control study (cases and selected controls drawn from a parent cohort study of a screening program in British Columbia), Caussy et al. [41] observed no difference in the HPV positivity rates from 47 biopsy samples of women with CIN who later progressed to invasive cancer and 94 CIN patients (controls) in whom no progression occurred. Use of in situ hybridization in paraffin blocks could have decreased the rate of detection in this study. These data do suggest that other factors may be relevant in the progression from CIN to cervical cancer. Murthy et al. [42] reached the opposite conclusion in a nested case-control study within a cohort of dysplasia women. Of 64 women suffering progression from dysplasia to CIN, the prevalence of HPV 16/18 was 68% compared to 27% among nonprogressing dysplastic cases.

The prospective study of 241 STD clinic attendees with negative Pap smears has demonstrated that women who become infected with HPV have an increased risk of developing CIN and cervical cancer (RR: 11; 95% CI: 4–33) [43]. It is not clear whether low-grade lesions actually progress to high-grade lesions. It has been suggested that high-grade lesions may arise from normal-appearing epithelium at the internal margin of low-grade lesions [44].

Two major cohorts continue to be followed [43, 45]. The largest is that of Schiffman et al. [45] with the follow-up of 18,000 cytologically negative women. Important findings to support the etiologic role of HPV in cervical cancer will emanate from this study.

Other Anogenital Cancers Related to HPV

The epithelial structures of the anorectal junction joining squamous and columnar epithelia are quite similar to the transformation zone of the uterine cervix. It is quite possible, therefore, that the anal mucosa could support the development and eventual progression of HPV-related dysplastic lesions similar to those occurring in the cervix.

Until recently, anal carcinoma had been a relatively rare tumor; it occurs most frequently in women and individuals over the age of 50 years. Even before the AIDS epidemic, investigators had questioned a potential relationship between male homosexual behavior, i.e. receptive anal intercourse, and the incidence of anal cancer. Dahling et al. [46] reported that 24.4% of the men with anal cancer in their study had never been married compared to only 7.8% of the men with colon and rectal cancer. This relationship was never documented among women with anal cancer. The precise mechanism of this association was not addressed; however, the association itself was rapidly supported by other investigators [47, 48].

In a later case-control study by Dahling et al. [49] of 148 men with anal cancers and 166 controls, it was demonstrated that receptive anal intercourse in men is a risk factor for anal cancer (RR: 33.1; CI: 4.0–272). In women anal intercourse was only weakly associated with anal cancer (RR: 1.8; CI: 0.7–4.2). Dahling et al. [49] determined that the squamous cell cancers were associated with genital warts, suggesting that HPV has an etiologic role in anal cancers. Palefsky et al. [50] determined that 85% of anal cancers are HPV positive using PCR. Lesions infected with HPV 16/18 were neoplastic in histology; those infected with HPV 6/11 were morphologically judged to be benign condylomata.

Melbye et al. [51] characterized risk factors for the presence of HPV 16/18 and examined their relationship to anal intraepithelial neoplasia (AIN) in 120 Danish homosexual men. Detection of HPV was significantly associated with HIV positivity and level of immunosuppression as measured by T lymphocyte subset markers. No association was found between HPV infection and the number of sexual partners, smoking, alcohol and illegal drug intake. History of anal condyloma in the past year was correlated with HPV. This association was specific to HPV 16, 18, 31, 33 and 35. Anal lesions were present in 19% of the men and were strongly correlated with the presence of HPV. The authors concluded that HPV plays a central role in the development of AIN in homosexual men [51].

More recent work by Palefsky et al. [52] has provided data to strongly support their contention that the spectrum of HPV presentation is changing [52, 53]. These investigators suggest that the incidence of AIN in symptomatic HIV-positive men is rapidly increasing and is approaching the commonality of genital condyloma. In addition, they offer evidence supporting a sharp increase in anal cancers among the HIV-positive males in pandemic areas of the US. Palefsky et al. [52] followed 97 HIV-positive men (group IV) and detected HPV DNA in 54% and abnormal anal cytology in 39%. The presence of HPV was significantly associated with abnormal cytology (RR: 4.6; p = 0.003). AIN was detected in 15% of their sample. T4 counts were found to be significantly decreased in HIV-positive men with abnormal as compared to normal cytology.

Beckmann et al. [54] have also studied specimens from 70 patients with squamous neoplasia of the anal skin or mucosa: 23 were positive for HPV. In addition, some malignant tumors of the anus in both men and women were associated with HPV infection.

Palefsky et al. [55] followed a cohort of 37 homosexual men with group IV HIV disease for an average follow-up of 17 months. During the study, the incidence of AIN increased from 27 to 65%. High-grade lesions increased from 0 to 16%. Detection of anal HPV increased from 60 to 89%. This study established the significant risk of AIN among homosexual males.

In 1993, Kiviat et al. [56] quantified the level of HIV-specific immunosuppression in association with AIN and HPV, and examined the mechanisms of the association. In this cross-sectional study of gay men in an HIV-screening clinic, anal HPV DNA was detected by Southern transfer hybridization in 55% of the 285 HIV-positive men compared to 23% of the 204 negative men (RR: 4.0; CI: 2.7–6.2). These rates of HPV infection increased to 92% among the HIV-seropositive men and 78% in the seronegative men with PCR testing. Twenty-six percent of those positive for HIV and only 8% of those who were negative had AIN (RR: 5.6; CI: 3–10.5). High-grade lesions occurred in 4% and 0.5% of the two groups, respectively.

Additional evidence is available to suggest that HPV is a risk factor for penile cancer. In a population-based case-control study in Washington state and British Columbia, 110 men with penile cancer occurring between 1979 and 1990 were compared to 355 controls. Tumor tissue was examined from 67 of the cases. No circumcision at birth, smoking and history of genital warts were related to penile cancer, but 49% of the tumors were positive for HPV with 70% of these being specific for HPV 16/18 [57].

Brandsma et al. [58] demonstrated a conspicuous and specific association of HPV with verrucous carcinoma of the larynx in tissue specimens of a small group of patients. HPV sequences similar but not identical to HPV 16 were identified as was the presence of koilocytosis. Although a pathogenic involvement of HPV seemed feasible, the presence of HPV as a passenger virus was not ruled out.

Current research suggests that there may exist an association between HPV and oral mucosal lesions. Both oral and cervical endothelium are exposed to carcinogenic substances and microorganisms, some of which may be carcinogenic. Each site has premalignant lesions which progress from mild dysplasia to carcinoma in situ and invasive carcinoma. Reports regarding the presence of HPV genotypes within oral tumors by molecular hybridization techniques do appear in the literature [59–61]; however, the findings are not conclusive.

HPV-Related Cancers Among Immunosuppressed Individuals

Numerous accounts of increased rates of HPV-related lesions among immunosuppressed individuals have been reported in the literature. Although observational studies cannot directly address basic mechanisms of an association, the study of specific populations in which associations appear to be magnified often provides insights for the generation of hypotheses and for future studies. The initial suggestion of an association between immunosuppression and HPV infection emanated from the coincident occurrence of chronic, extensive papillomatous lesions due to certain HPV types such as 5 and 8, and the development of squamous cell skin cancers in 30–100% of patients with epidermodysplasia verruciformis, a condition of cell-mediated immune defects [62–64]. The impact of iatrogenic immunosuppression on susceptibility to HPV has been examined in numerous studies of females who have undergone renal transplantation. Halpert et al. [65] demonstrated a 7-fold increase in risk of HPV-induced lesions and a 17-fold increase in the incidence of cervical cancer among females receiving immunosuppressant therapy after renal allograft [65]. Other investigators have reported a 5-fold greater risk of cervical cancer among such women compared to age-matched controls [66]. Similar increases in risk of cervical carcinoma have been confirmed in women receiving chemotherapy and radiation for Hodgkin's disease [67]. Increased rates of HPV-induced lesions have also been demonstrated in otherwise healthy women during the transient immunosuppression of pregnancy. These lesions tend to rapidly regress postpartum [68]. These clinical observations have sparked strong suspicions that the host immune system may be a vital factor in the etiologic cascade leading to malignancy and that the expression and detection of HPV may be modulated by the host immune response.

Immunologic Evidence of HPV Association

Historically, data related to the immmunologic response to HPV have been inconsistent and difficult to interpret due to variability of assays, inadequate assessment of HPV type specificity and inadequate characterization of the study populations with respect to HPV infection [69]. Recently, investigations into the humoral immune response have begun to formulate consistent findings. Evidence suggests that antibodies to epitopes on HPV 16 E7 may be markers for HPV-16-related invasive cancers [70], and similarly HPV 18 E7 has been associated with HPV-18-associated cervical cancers [71]. The use of these biologic markers in well-designed epidemiologic studies can contribute another dimension to establishing a causal association between HPV and cervical cancer.

Among women evaluated in Spain and Columbia, Müller et al. [72] demonstrated significantly increased antibody activity among those with cervical cancer (49%) compared to those with CIN (24%) and age-matched controls (17%; $p < 0.001$). A recent preliminary report of 137 women diagnosed with cervical cancer suggested that an HPV-specific humoral immune response may increase with the cervical cancer stage and decrease following tumor therapy [73]. Very little data are available regarding cell-mediated responses with HPV infection. Cell-mediated immunity among patients with epidermodysplacia verruciformis is characterized by decreased cytotoxicity and decreased T helper cells [74, 75]. Decreased cellular immune responses have also been demonstrated in individuals with chronic warts, suggesting a persistent failure of the cellular immune system to respond to HPV infection [76, 77].

Conclusion

Data from well-designed, well-controlled epidemiologic studies addressing the association of HPV and cervical cancer are mounting. Each new contribution must be critically reviewed and objectively judged regarding the strength of its findings and its support of HPV causality. As discussed, the strength of the association of HPV and cervical cancer is relatively high. Risk is estimated to be from a minimum of 2-fold to extremely high magnitudes over non-HPV-infected risks. Although detection techniques still introduce variability among studies and laboratories, recent studies are providing consistency in demonstrating positive results. Scientific plausibility of HPV as the etiologic agent in cervical cancer is well supported by examining the risk factors related to HPV and cervical cancer, as well as the extensive molecular studies that appear to demonstrate the ability of HPV to transform normal cells into malignant cells. Although HPV is not definitively necessary for cervical cancer development, there remains only a very small proportion of cervical tumors which test negative for HPV. This proportion is likely to continue to decrease as new types of the HPV virus are identified. Since many HPV-infected females have normal cytologic exams, HPV is clearly not sufficient for the development of cervical cancer; additional factors must be present to allow tumor growth. Further consideration must be given to fixed cofactors such as demographics, genetics and diet as well as variable cofactors, including pregnancy, hormone exposure, smoking and STD. Ongoing and future cohort studies are an important link in confirming this association. As consistent data from such studies continue to mount, the etiologic association of HPV and cervical cancer will be more firmly established and significant light will be shed on the role of cofactors in this association. Evidence of a biologic gradient has been demonstrated regarding increasing HPV infection rates associated with increasing levels of dysplasia, increasing virus load signal with increasing dysplasia and, lastly, preliminary data suggesting increasing antibody levels with increasing stage of cervical cancer.

The evidence from the laboratory and the observational studies, at first glance, is compelling; however, additional epidemiologic studies to clearly define the role of confounding in causation will contribute significantly to our understanding. The tradition of understanding cause and effect, including specification of the mechanism by which the effect is produced, most often ends in an opinion or judgement about causality rather than a claim of proof. Certainly, full knowledge of etiologic mechanisms is not necessary in order to develop successful approaches to disease prevention and control. Given the strong role of HPV in cervical cancer our hopes for prevention and treatment approaches may be fulfilled in the not too distant future. However, until the exact mechanism of HPV causation in cancer can be clearly defined, all investigators must be open-minded in our resolve to weigh each new scientific contribution and continually examine and question the complex, multifactorial interactions inherent in the mystical relationship between HPV and carcinogenesis.

References

1 Mausner JS, Kramer S: Epidemiology – An Introductory Text. Philadelphia, Saunders, 1985.
2 Rothman KJ: Modern Epidemiology. Boston, Little, Brown, 1986, p 11.
3 MacMahon B, Pugh TF: Epidemiology – Principles and Methods. Boston, Little, Brown, 1970, p 23.
4 Hennekens CH, Buring JE: Epidemiology in Medicine. Boston, Little, Brown, 1987, p 39.
5 Hill AB: The environment and disease: Association or causation? Proc R Soc Med 1965;58: 295–300.
6 Silverberg E, Boring CC, Squires TS: Cancer statistics 1990. CA 1990;40:9–26.
7 Brinton LA: Epidemiology of cervical cancer; in Munoz N, Bosch FX, Shah KV, Meheus A (eds): The Epidemiology of Human Papillomavirus and Cervical Cancer. Lyon, ARC, 1992, pp 3–23.
8 Brinton LA, Hamman RF, Huggins GR: Sexual and reproductive risk factors for invasive squamous cell cervical cancer. J Natl Cancer Inst 1987;79:23–29.
9 Slattery ML, Overall JC, Abbott TM, French TK, Robison LM, Gardner J: Sexual activity, contraception, genital infections, and cervical cancer: Support for a sexually transmitted disease hypothesis. Am J Epidemiol 1989;130: 248–258.
10 Jones CJ, Brinton LA, Hamman RF, Stolley PD, Lehman HF, Levine RS, Mallin K: Risk factors for in situ cervical cancer: Results from a case-control study. Cancer Res 1990;50: 3657–3662.
11 Taylor RS, Carroll BE, Lloyd JW: Mortality among women in 3 Catholic religious orders with special references to cancer. Cancer 1959; 12:1207–1223.
12 Ley C, Bauer HM, Reingold A, Schiffman MH, Chambers JC, Tashiro CJ, Manos MM: Determinants of genital human papillomavirus infection in young women. J Natl Cancer Inst 1991;83:997–1003.
13 Schiffman MH, Haley NJ, Felton JS, Andrews AW, Kaslow RA, Lancaster WD, Kurman RJ, Brinton LA, Lannom LB, Hoffmann D: Biochemical epidemiology of cervical neoplasia: Measuring cigarette smoke constituents in the cervix. Cancer Res 1987;47:3886–3888.
14 Franco EL: Viral etiology of cervical cancer: A critique of the evidence. Rev Infect Dis 1991; 13:1195–1206.
15 Vonka V, Kanka J, Roth Z: Herpes simplex type 2 virus and cervical neoplasia. Adv Cancer Res 1987;48:149–191.
16 Meisels A, Fortin R: Condylomatous lesions of the cervix and vagina. I: Cytologic patterns. Acta Cytol 1976;20:505–509.
17 zur Hausen H: Human papillomaviruses and their possible role in squamous cell carcinoma. Curr Top Microbiol Immunol 1977;78:1–30.
18 Gissmann L, de Villiers EM, zur Hausen H: Analysis of human genital warts (condyloma acuminata) and other genital tumors for human papillomavirus type 6 DNA. Int J Cancer 1982;29:143–146.
19 zur Hausen H: Papillomaviruses as carcinoma viruses. Adv Viral Oncol 1989;8:1–26.
20 Munoz N, Bosch FX: HPV and cervical neoplasia: Review of case-control and cohort studies; in Munoz N, Bosch FX, Shah KV, Meheus A (eds): The epidemiology of Cervical Cancer and Human Papillomavirus. Lyon, IARC, 1992, pp 251–261.
21 Kulski JK, Demeter T, Mutavdzic S, Sterrett GF, Mitchell KM, Pixley E: Survey of histologic specimens of human cancer for human papillomavirus types 6/11/16/18 by filter in situ hybridization. Am J Clin Pathol 1990;94: 566–570.
22 Yoshikawa H, Kawana T, Kitagawa K, Mizuno M, Yoshikura H, Iwamoto A: Detection and typing of multiple genital human papillomaviruses by DNA amplification with consensus primers. Jpn J Cancer Res 1991;82:524–531.
23 Peters RK, Thomas D, Hagan DG, Mack TM, Henderson BE: Risk factors for invasive cervical cancer among Latinas and non-Latinas in Los Angeles County. J Natl Cancer Inst 1986; 77:1063–1077.
24 La Vecchia C, Franceschi S, Decarli A: Sexual factors, venereal diseases, and the risk of intraepithelial and invasive cervical neoplasia. Cancer 1986;58:935–941.
25 Kjaer SK, de Villiers EM, Dahl C, Engholm G, Bock JE, Vestergaard BF, Lynge E, Jensen OM: Case-control study of risk factors for cervical neoplasia in Denmark. I: Role of the 'male factor' in women with one lifetime sexual partner. Int J Cancer 1991;48:39–44.
26 Munoz N, Bosch FX, Sanjose S, Tafur L, Izarzugaza I, Gili M, Viladiu P, Navarro C, Martos C, Ascunce N, Gonzalex LC, Kaldor JM, Guerrero E, Lorincz A, Santamaria M, Alonso de Ruiz P, Aristizabal N, Shah K: The causal link between human papillomavirus and invasive cervical cancer: A population-based case-control study in Colombia and Spain. Int J Cancer, in press.
27 de Villiers EM, Wagner D, Schneider A, Wesch H, Miklaw H, Wahrendorf J, Papendick U, zur Hausen H: Human papillomavirus infections in women with and without abnormal cervical cytology. Lancet 1987;ii:703–706.
28 Lorincz AT, Reid R, Jenson AB, Greenberg MD, Lancaster W, Kurman RJ: Human papillomavirus infection of the cervix: Relative risk associations of 15 common anogenital types. Obstet Gynecol 1992;79:328–337.
29 Kiviat NB, Koutsky LA, Paavonen JA, Galloway DA, Critchlow CW, Beckmann AM, McDougall JK, Peterson ML, Stevens CE, Lipinski CM, Holmes KK: Prevalence of genital papillomavirus infection among women attending a college student health clinic or a sexually transmitted disease clinic. J Infect Dis 1989;159:293–302.
30 Manos MM, Schiffman MH, Bauer HM, Robinson EK, Glass AG, Cadell DM, Kurman RJ: HPV and cervical neoplasia: A prevalent case-control study (abstract). Papillomavirus Workshop, Seattle 1991, p 108.
31 Morrison EA, Ho GY, Vermund SH, Goldberg GL, Kadish AS, Kelley KF, Burk RD: Human papillomavirus infection and other risk factors for cervical neoplasia: A case-control study. Int J Cancer 1991;49:6–13.
32 Reeves WC, Brinto LA, Garcia M, Bnrenes MM, Herrero R, Gaitan E, Tenorio F, de Britton RC, Rawls WE: Human papillomavirus infection and cervical cancer in Latin America. N Engl J Med 1989;320:1436–1441.
33 Cuzick J, Singer A, De Stavola BL, Chomet J: Case-control study of risk factors for cervical intraepithelial neoplasia in young women. Eur J Cancer 1990;26:684–690.
34 Barnes W, Delgado G, Kurman RJ, Petrilli ES, Smith DM, Ahmed S, Lorincz AT, Temple GF, Jenson AB, Lancaster WD: Possible prognostic significance of human papillomavirus type in cervical cancer. Gynecol Oncol 1988;29:267–273.
35 Lancaster WD, Castellano C, Santos C, Delgado G, Kurman RJ, Jenson AB: Human papillomavirus deoxyribonucleic acid in cervical carcinoma from primary and metastatic sites. Am J Obstet Gynecol 1986;154:115–119.
36 Walboomers JM, Fokke HE, Polak M, Volers H, Houthoff HJ, Barents J, van der Noordaa J, ter Schegget J: In situ localization of human papilloma virus type 16 DNA in a metastasis of an endocervical adenocarcinoma. Intervirology 1987;27:81–85.
37 Sutton GP, Stehman FB, Ehrlick CE, Roman A: Human papillomavirus deoxyribonucleic acid in lesions of the female genital tract: Evidence for type 6/11 in squamous carcinoma of the vulva. Obstet Gynecol 1987;70:564–568.
38 Mitchell H, Drake M, Medley G: Prospective evaluation of risk of cervical cancer after cytological evidence of human papillomavirus infection. Lancet 1986;i:573–575.
39 Campion MJ, McCance DJ, Cuzick J, Singer A: Progressive potential of mild cervical atypia: Prospective cytological, colposcopic, and virological study. Lancet 1986;ii:237–240.
40 Schneider A, Sawada I, Gissmann L, Shah K: Human papillomaviruses in women with a history of abnormal Papanicolaou smears and in their male partners. Obstet Gynecol 1987;69: 554–562.
41 Caussy D, Marrett LD, Worth AJ, McBride M, Rawls WE: Human papillomavirus and cervical intraepithelial neoplasia in women who subsequently had invasive cancer. Can Med Assoc J 1990;142:311–317.
42 Murthy NS, Sehgal A, Satuanarayanan L, Das DK, Singh V, Das BC, Gupta MM, Mitra AB, Luthra UK: Risk factors related to biological behavior of precancerous lesions of the uterine cervix. Br J Cancer 1990;61:732–736.

43 Koutsky LA, Holmes KK, Critchlow CW, Stevens CE, Paavonen J, Beckmann AM, DeRouen TA, Galloway DA, Vernon D, Diviat NB: A cohort study of the risk of cervical intraepithelial neoplasia grade 2 or 3 in relation to papillomavirus infection. N Engl J Med 1992;327: 1272–1277.
44 Kiviat NB, Critchlow CW, Kurman RJ: Reassessment of the morphological continuum of cervical intraepithelial lesions: Does it reflect different stages in the progression to cervical carcinoma? in Munoz N, Bosch FX, Shah KV, Meheus A (eds): The Epidemiology of Human Papillomavirus and Cervical Cancer. Lyon, IARC, 1992, pp 59–66.
45 Schiffman MH, Bauer HM, Glass AG, Cadell DM, Zhang TY, Kurman RJ, Manos MM: Preliminary results from a prospective study of HPV infection and incident cervical neoplasia. Papillomavirus Workshop, Seattle, 1991, p 107.
46 Dahling JR, Weiss NS, Klopfenstein LL, Cochran LE, Chow WH, Daifuku R: Correlates of homosexual behavior and the incidence of anal cancer. JAMA 1982;247:1988–1990.
47 Peters RK, Mack TM: Patterns of anal carcinoma by gender and marital status in Los Angeles County. Br J Cancer 1983;48:629–636.
48 Austin DF: Etiologic Clues from Descriptive Epidemiology: Squamous Carcinoma of the Rectum or Anus. Bethesda, National Cancer Institute, 1982, pp 89–90.
49 Dahling JR, Weiss NS, Hislop TG, Maden C, Coates RJ, Sherman KJ, Ashley RL, Beagrie M, Ryan JA, Corey L: Sexual practices, sexually transmitted diseases, and the incidence of anal cancer. N Engl J Med 1987;317:973–977.
50 Palefsky JM, Holly EA, Gonzales J, Berline J, Ahn DK, Greenspan JS: Detection of human papillomavirus DNA in anal intraepithelial neoplasia and anal cancer. Cancer Res 1991; 51:1014–1019.
51 Melbye M, Palefsky J, Gonzales J, Ryder LP, Nielsen H, Bergmann O, Pindborg J, Biggar RJ: Immune status as a determinant of human papillomavirus detection and its association with anal epithelial abnormalities. Int J Cancer 1990;46:203–206.
52 Palefsky JM, Gonzales J, Greenblat RM, Ahn DK, Hollander H: Prevalence of anal intraepithelial neoplasia and anal papillomavirus infection among immunosuppressed homosexual males with group IV HIV disease. JAMA 1990; 263:2911–2916.
53 Biggar RJ, Burnett W, Mikl J, Nasca P: Cancer among New York men at risk of acquired immunodeficiency syndrome. Int J Cancer 1989;43:979–985.
54 Beckmann AM, Dahling JR, Sherman KJ, Maden C, Miller BA, Coates RJ, Kiviat NB, Myerson D, Weiss NS, Hislop G, Beagrie M, McDougall JK: Human papillomavirus infection and anal cancer. Int J Cancer 1989;43: 1042–1049.
55 Palefsky JM, Holly EA, Gonzales J, Lamoborn K, Hollander H: Natural history of anal cytologic abnormalities and papillomavirus infection among homosexual men with group IV HIV disease. J Acquir Immune Defic Syndr 1992;5:1258–1265.
56 Kiviat NB, Critchlow CW, Holmes KK, Kuypers J, Sayer J, Dunphy C, Surawicz C, Kirby P, Wood R, Dahling JR: Association of anal dysplasia and human papillomavirus with immunosuppression and HIV infection among homosexual men. AIDS 1993;7:43–49.
57 Maden C, Sherman KJ, Beckmann AM, Hislop TG, Teh CZ, Ashley RL, Dahling JR: History of circumcision, medical conditions, and sexual activity and risk of penile cancer. J Natl Cancer Inst 1993;85:19–24.
58 Brandsma JL, Steinberg BM, Abramson AL, Winkler B: Presence of human papillomavirus type 16 related sequences in verrucous carcinoma of the larynx. Cancer Res 1986;46:2185–2188.
59 De Villiers EM, Newman C, Le JY, Weidaner H, zur Hausen H: Infections of the oral mucosa with defined types of human papillomaviruses. Med Microbiol Immunol 1987;174:287–293.
60 Lookingbill DP, Dreider JW, Howett MK, Olmstead PM, Conners GH: Human papillomavirus type 16 in bowenoid papillomas, intraoral papillomas and squamous cell carcinomas of the tongue. Arch Dermatol 1987;123: 363–368.
61 Adler-Stroth K, Newland JR, Tessin BA, Yendall WA, Shillte EJ: Human papillomavirus type 2 DNA in oral verrucous carcinoma. J Oral Pathol 1986;15:472–474.
62 Glinski W, Obalek S, Jablonska S, Orth G: T-cell defect in patients with epidermodysplasia verruciformis due to human papillomavirus types 3 and 5. Dermatologica 1981;162:141–147.
63 Haftek M, Jablonska S, Orth G: Specific cell-mediated immunity in patients with epidemodysplasia verruciformis and plane warts. Dermatologica 1985;170:213–220.
64 Jablonska S, Dabrowski J, Jakubowicz K: Epidermodysplasia verruciformis as a model in studies on the role of papillomaviruses in oncogenesis. Cancer Res 1972;32:583.
65 Halpert R, Fruchter RG, Sedlis A, Butt K, Boyce JG, Stillman FH: Human papillomavirus and lower genital neoplasia in renal transplant patients. Obstet Gynecol 1986;68:251–257.
66 Penn I: Cancers of the anogenital region in renal transplant recipients: Analysis of 65 cases. Cancer 1986;8:611.
67 Katz RL, Veanattukalathil S, Weiss KM: Human papillomavirus infection and neoplasia of the cervix and anogenital region in women with Hodgkin's disease. Acta Cytol 1987;31:845–854.
68 Schneider V, Hotz M, Gissmann L: Increased prevalence of human papillomavirus in the lower genital tract of pregnant women. Int J Cancer 1987;40:198–201.
69 Palefsky J: Human papillomavirus infection among HIV-infected individuals. Hematol Oncol Clin North Am 1991;5:357–370.
70 Mann VM, Lood De Lao S, Brenes M, Brinton LA, Rawls JA, Green M, Reeves WC, Rawls WE: Occurrence of IgA and IgG antibodies to select peptides representing human papillomavirus type 16 among cervical cancer cases and controls. Cancer Res 1990;50:7815–7819.
71 Bleul C, Müller M, Frank F, Gausepohl H, Kildovsky U, Mgaya HBN, Luande J, Pawlita M, ter Muelen J, Viscidi R, Gissmann L: Human papillomavirus type 18 E6 and E7 antibodies in human sera: Increased prevalence in cervical cancer patients. J Clin Microbiol 1991;29: 1579–1588.
72 Müller M, Viscidi RP, Sun Y, Guerrero E, Hill PM, Shah F, Bosch FX, Munoz N, Gissmann L, Shah KV: Antibodies to HPV-16 E6 and E7 proteins as markers for HPV-16-associated invasive cervical cancer. Virology 1992;187:508–514.
73 Fisher SG, Benitez-Bribiesca L, Nindl I, Gissman L: The association of human papillomavirus type 16 E6 and E7 antibodies with stage of cervical cancer, submitted.
74 Kaminski M, Pawinska M, Jablonska S, Szmurlo A, Majewski S, Orth G: Increased natural killer cell activity in patients with epidermodysplasia verruciformis. Arch Dermatol 1985;121:84.
75 Majewski S, Malejczyk J, Jablonska S, Misiewicz J, Rudnicka L, Obalek S, Orth G: Natural cell-mediated cytotoxicity against various target cells in patients with epidermodysplasia verruciformis. J Am Acad Dermatol 1990;22: 423.
76 Morison WL: Viral warts, herpes simplex, and herpes zoster in patients with secondary immune deficiencies and neoplasms. Br J Dermatol 1975;92:625.
77 Cauda R, Tyring SK, Grossi CE, Tilden AB, Hatch KD, Sams WM Jr, Baron S, Whitley RJ: Patients with condyloma accuminata exhibit decreased interleukin-2 and interferon gamma production and depressed natural killer activity. J Clin Immunol 1987;7:304–311.

Intervirology 1994;37:226–230

Lex M. Cowsert

Department of Infectious Disease, ISIS Pharmaceuticals, Carlsbad, Calif., USA

Treatment of Papillomavirus Infections: Recent Practice and Future Approaches

Key Words
Papillomavirus
Therapeutics
Antisense
Condyloma

Introduction

Human papillomavirus (HPV) infection is associated with epithelial neoplasms ranging from benign common warts to invasive carcinoma of the cervix. While HPV can infect many anatomical sites, it is infection of the anogenital region that is most important. The significance of genital HPV infection is derived from its sexual mode of transmission, near epidemic proportions [1], and the clear association of certain HPV types with the development of cervical dysplasia and progression of dysplasia to cervical cancer [2]. Because of its sexual mode of transmission, genital HPV infection is particularly prevalent among the young sexually active population. For example, one study documented that 46% of women attending a university health center for routine gynecological examination were infected with HPV [3]. The enormity of the problem is suggested by the fact that in 1990 the Centers for Disease Control (CDC) estimated that 800,000 new cases of HPV infection were diagnosed in the United States.

Despite the prevalence of genital HPV infection and the association with malignant disease, there is no proven antiviral therapy for its treatment. The goal of current therapeutic strategies is the removal of exophytic warts and the elimination of signs and symptoms, not eradication of viral DNA. While nearly all genital warts can be eliminated by such treatments as cryotherapy, electrodesiccation, or surgical removal, recurrence rates are generally unacceptable. This is probably due to residual virus in adjacent normal tissue.

In general, chemotherapeutic approaches to control HPV infection suffer from a lack of specificity. Papillomaviruses do not encode their own DNA polymerase and must therefore rely heavily upon cellular machinery for replication. This dependence on cellular pathways for replication may explain, in part, the stringent tissue tropism displayed by papillomaviruses. This tight association with cellular pathways and lack of viral polymerases makes it difficult to identify chemotherapeutic agents with appropriate selectivity.

Many studies have been carried out to compare the relative efficacies of available treatments for genital HPV infection. Efficacies range from 22 to 94% for clearance of exophytic genital warts. Recurrence rates are typically high, usually in the range of 25% within 3 months. Effectiveness of the treatments is only a partial explanation for the wide range of success rates and high recurrence rates. Other factors, including size, age, and anatomical location of the lesion as well as the skill and aggressiveness of the practitioner contribute significantly to the overall success rate of any therapy. Difficulties in controlling for these factors in clinical studies may lead to over- or underestimations of the effectiveness of a particular therapy.

The CDC have recently published their guidelines for the treatment of external genital/perianal warts, vaginal warts, urethral meatus warts, and anal warts [4]. Cryo-

Lex M. Cowsert, PhD
Department of Infectious Disease
ISIS Pharmaceuticals
2280 Faraday Avenue
Carlsbad, CA 92008 (USA)

therapy was listed as the preferred therapy for genital warts. The CDC suggest that the selection of therapy be guided by patient preference and consideration be given to expense, toxicity, and potential for scarring. Other recommended therapies specifically included podophyllin, podophyllotoxin, trichloroacetic acid, and electrodesiccation or electrocautery. Carbon dioxide laser and conventional surgery were recommended for the management of extensive or refractory disease. Interferon was not recommended because of cost, high frequency of adverse side effects, and poor response. 5-fluorouracil (5-FU) was not recommended for treatment of genital warts.

Current Therapies

Cryotherapy

Cryotherapy using liquid nitrogen or cryoprobe is the treatment of choice for genital warts according to the CDC. Cryotherapy works by destroying treated tissue and stimulating the immune response [5, 6]. Liquid nitrogen is applied using a cotton-tipped applicator or spray delivery. Lesions and a margin of 1–2 mm should be completely frozen. The size and thickness will determine the number of treatments required for resolution of the wart. Side effects of cryotherapy include pain, discomfort, edema, necrosis, ulceration, and blister formation. The number and severity of the side effects increase with the number of treatments.

The majority of cryotherapy patients exprience resolution of their warts with 3 treatments or less [7–9]. The success rate for cryotherapy is in the range of 70% [7–9]. The number of warts treated and anatomical location correlate with success rates. For example, anal or perianal warts are more resistant to cryotherapy often requiring more than three treatments for complete resolution. The number of warts treated can reduce the effectiveness of cryotherapy. Patients with 1–3 warts required on average 1.37 treatments while patients with more than 3 warts required 2.33 treatments.

Cryotherapy is also the treatment of choice for women in the second and third trimester of pregnancy [10]. Cryotherapy is also an efficacious treatment for cervical intraepithelial neoplasia (CIN). 94% of women treated with cryotherapy had normal cytology and colposcopic findings at 1-year follow up and 89% of women available for follow up at 10 years had no evidence of CIN [11].

Surgical Treatments

In general, surgical approaches to the treatment of genital warts, such as excision, electrodesiccation, electrocautery, and CO_2 laser, have higher success, require fewer treatment sessions, and patients have fewer recurrences than with cryotherapy or medical approaches. Certain factors, such as lesion size, persistence, anatomical location, or resistance to other therapies, may indicate a surgical approach. However, surgical techniques do have drawbacks. Side effects can include pain and scarring and the cost/benefit needs to be carefully considered.

Many studies have been carried out to determine the efficacy of various surgical therapies relative to each other as well as to cryotherapy and medicinal approaches. A randomized prospective study comparing surgical excision to podophyllin for first-episode perianal condyloma demonstrated success rates of 93 and 76%, respectively [12]. Patients treated by surgical excision experienced significantly fewer recurrences than those in the podophyllin treatment group: 29 compared to 65% at 12 months. When compared in the same study, electrodesiccation was more effective than cryotherapy at 3 months follow up (91 versus 62%) [13]. On average, fewer treatments were required for clearance by electrodesiccation (1.4) than by cryotherapy (2.6).

Podophyllin and Podophyllotoxin

Podophyllin is a crude extract prepared from the roots of the may apple plant, *Podophyllin peltatum*, and is one of the more commonly used treatments for genital condyloma [14, 15]. Advantages of podophyllin include ease of use, patient self-administration, and low cost. However, success rates for podophyllin vary greatly. Podophyllin does have significant local and systemic side effects that increase in incidence and severity with increased treatment surface area thus limiting its use for widespread disease.

Podophyllotoxin, the active ingredient in podophyllin, was approved by the Food and Drug Administration (FDA) in 1991 under the name of Condylox (Oclassen Pharmaceuticals, San Rafael, Calif., USA) for the treatment of genital condyloma. Podophyllotoxin is more active than podophyllin (94% effective compared to 29%) with recurrences of 23 versus 38% [16]. Podophyllotoxin is sufficiently safe and effective that it can be patient-administered for the treatment of genital condyloma. In one study, self-administered podophyllotoxin produced a greater reduction in wart area than placebo (82 versus 4%) [17]. The range in reported success rates using podophyllotoxin is 100% with 23% recurrence [18] to 45% with 40% recurrence [17].

Chemotherapy

Aggressive chemotherapy using a variety of compounds – 5-FU [19], bleomycin [20], and methotrexate [21] – with antiproliferative activity has been used for the treatment of severe HPV-induced disease. The therapeutic potential for these compounds was first explored in anticancer strategies where selectivity was based on their preferential effects on dividing cells. However, current evidence suggests that cells in papillomas do not divide more rapidly than normal cells in vitro [22] and the mitotic index for papillomas is no higher than in surrounding normal tissue in vivo [23].

5-FU has been used for many years for the treatment of HPV-associated disease [21]. It is not FDA-approved for this purpose. 5-FU has been shown to be teratogenic in animals and should not be used in pregnant women [24]. 5-FU acts by inhibition of DNA and RNA synthesis and is more readily absorbed by abnormal than normal skin which may afford some selectivity [19]. 5-FU is also immunostimulatory and elicits a local hypersensitivity that augments its cytotoxic effects. 5-FU has been successfully used for the treatment of vaginal condylomas and vaginal intraepithelial neoplasia [19]. The highest success rates are observed for the treatment of multiple small nonkeratinizing lesions. In specific instances, 5-FU can achieve excellent results, particularly in refractory disease, either as a primary or an adjunctive therapy. Further, 5-FU may have significant value as a prophylactic treatment against future recurrences in immunocompetent [25] and immunosuppressed [26] patients.

Prospects for the Future

The life cycle of papillomaviruses is intimately associated with the biology of the host cell and is dependent upon host functions for many aspects of the life cycle. It therefore becomes difficult to imagine how traditional classes of antivirals could provide the selectivity necessary for therapeutic effects. Despite the lack of an in vitro system permissive to papillomavirus replication, the genes responsible for many critical viral functions, such as viral DNA replication [27, 28], perturbation of differentiation [29], and transformation [30], have been identified. Thus it now becomes possible to develop rationales to target specific viral genes or their products for therapeutic benefit.

Over the past decade there have been enormous advances in our understanding of the biology and epidemiology of papillomaviruses. One of the major hindrances to the development of papillomavirus-specific antivirals has been the lack of suitable in vitro systems permissive for viral replication. Recent technical advances, however, have allowed the development of in vitro model systems permissive to HPV replication [31, 32]. These models while technically difficult may be adapted to screening drug compounds for antiviral activity. Animal models that recapitulate the entire life cycle of HPV have also been developed [33, 34]. Together these models may provide the necessary tools for in vitro screening and preclinical in vivo evaluation of candidate HPV medical therapies.

New formulations and combinations of existing therapies may yield new effective therapies for genital warts. For example, an intralesional implant, MPI 5003, containing collagen, epinephrine and 5-FU is currently in phase III clinical trials (Matrix Pharmaceuticals, Menlo Park, Calif., USA). In phase II clinical trials, MPI 5003 showed significant success (65%) in clearing genital warts and a greater than 50% reduction in wart area in 78% of patients [35].

Another promising approach is the application of photodynamic therapy to the treatment of HPV-associated lesions. Hematoporphyin derivative (HPD) when activated by light will destroy tissue where it accumulates in sufficient concentrations. When administered systematically at 5 mg/kg to rabbits there is insufficient HPD accumulation in normal skin at 24 h to elicit the photodynamic response [36]. However, HPD is still present at therapeutic concentrations in papillomas at 48 h. It is this selective retention of HPD in papillomas compared to normal skin that makes this approach possible. Photodynamic therapy is currently being applied for the treatment of laryngeal papillomatosis with some modest success [37]. Much effort is being put into optimizing this promising new technique for the treatment of a life-threatening condition.

All current papillomavirus therapies are inadequate in that they fail to address the viral infection in a specific way. While many surgical approaches may achieve >90% success in initial response, they all suffer from unacceptably high recurrence rates. This is thought to be due in large part to persistence of HPV in normal adjacent tissue. The intimate relationship of HPV with its host cell makes it difficult to imagine success using traditional antiviral strategies. Therefore, new strategies aimed at inhibition of viral gene expression may provide new approaches for the development of HPV-specific therapeutics.

Antisense Therapies

The antisense approach has been successfully applied to control HPV gene expression and the results have been very encouraging. Antisense can be applied as two broad strategies. In its simplest form, genes can be cloned and expressed in reverse orientation resulting in expression of antisense RNA that is complimentary to the target RNA. This strategy has been applied in an effort to reverse the transformed phenotype of cells containing integrated high-risk HPV types. In preliminary studies, cells derived from a cervical carcinoma (C4-I) containing integrated HPV 18 were transfected with dexamethasone-inducible expression vectors transcribing antisense RNA from the E6/E7 region of HPV 18 [38]. In these experiments there was a direct correlation between antisense RNA expression and reduction in growth rate. Expression of antisense RNA also resulted in a reduction in the E7 protein compared to controls. The experiments provide evidence for a significant role for E6 and E7 from high-risk HPVs in maintaining the transformed phenotype.

The application of antisense RNA strategies to reversal of the transformed phenotype of cells derived from cervical cancers containing high-risk HPV has been extended in vivo. Cell lines derived from the above experiments were transplanted into nude mice. Lines containing E6/E7 antisense constructs, but uninduced, were found to be as tumorigenic as the parental cell line [39]. However, when antisense RNA expression was induced by introduction of dexamethasone to the drinking water, tumors failed to develop from cell lines expressing HPV 18 E6/E7 antisense RNA. These data extend the previous in vitro observations and strongly support the hypothesis that expression of E6 and E7 is required for maintenance of the transformed phenotype, and clearly demonstrate the potential of antisense strategies for the control of HPV-associated disease.

The second antisense approach is the use of oligonucleotides complementary to mRNA of critical viral genes. A number of reports on efforts to use E6/E7 antisense oligonucleotides to inhibit proliferation of cells derived from cervical carcinomas containing high-risk HPV types have been published [40–42]. In all cases, oligonucleotides were targeted to E6/E7 RNAs. Because of overlapping open reading frames and mRNA splicing patterns [43], it is difficult to make antisense oligonucleotides that would target E6 or E7 individually. In all cases, antiproliferative effects were observed, although in some instances, effects were modest due to nonspecific antiproliferative effects of the phosphorothioate chemistry used to stabilize the oligonucleotides [41].

Antisense strategies targeting replication functions of papillomaviruses have been developed for treating benign HPV infection. E2 has been shown to be necessary, together with E1, for efficient DNA replication initiating from an HPV ori [27, 44]. Using bovine papillomavirus as a model, it was observed that inhibition of E2 by antisense oligonucleotides was both sequence and dose dependent and resulted in failure to establish infection in vitro [45]. ISIS 2105, a phosphorothioate antisense oligonucleotide targeting E2 of HPV 6 and HPV 11 is currently in phase II clinical trials. Phase I clinical trials for ISIS 2105 demonstrated that phosphorothioate oligonucleotides can be safely repeatedly dosed intradermally. If this E2 antireplication strategy proves to be successful, antisense oligonucleotides could be developed to other HPV types, such as HPV 1, for the treatment of plantar warts. However, E2 has been shown to be a negative regulator of E6/E7 expression in high-risk HPV types [46]. It has been hypothesized that one function of integration of high-risk HPVs is to eliminate E2 repression and thus up-regulate expression of the oncoproteins E6/E7. Therefore, E2-targeted oligonucleotides may be inappropriate for the treatment of low-grade dysplasias containing episomal high-risk HPVs. Other targets such as E1 may be more appropriate for this indication.

With the positive results experienced in the adenosine deaminase gene replacement therapy and recent demonstration of efficacy of antisense oligonucleotides in the treatment of cytomegalovirus retinitis in AIDS patients [47], genetically based strategies have moved from the theoretical to reality. There have been rapid advances in antisense/antigene technologies. Together with our increased understanding of papillomavirus biology, genetically based therapies seem poised to revolutionize HPV therapeutics.

References

1 Beutner KR, Becker TM, Stone KM: Epidemiology of human papillomavirus infections. Dermatol Clin 1991;9:211–218.

2 Schiffman MH, Bauer HM, Hoover RN, Glass AG, Cadell DM, Rush BB, Scott DR, Sherman ME, Kurman RJ, Wacholder S: Epidemiologic evidence showing that human papillomarivus infection causes most cervical intraepithelial neoplasia. J Natl Cancer Inst 1993;85:958–964.

3 Bauer HM, Ting Y, Greer CE, Chambers JC, Tashiro CJ, Chimera J, Reingold A, Manos MM: Genital human papillomavirus infection in female university students as determined by a PCR-based method. JAMA 1991;265:472–477.

4 Centers for Disease Control and Prevention: 1993 sexually transmitted diseases treatment guidelines. Morb Mortal Wkly Rep 1993;42:1–102.
5 Morison WL: Cell-mediated immune responses in patients with warts. Br J Dermatol 1975;93:553–556.
6 Morison WL: In vitro assay of immunity to human wart antigen. Br J Dermatol 1975;93:545–552.
7 Bashi SA: Cryotherapy versus podophyllin in the treatment of genital warts. Int J Dermatol 1985;24:535–536.
8 Balsdon MJ: Cryosurgery of genital warts (letter). Br J Vener Dis 1978;54:352–353.
9 Ghosh AK: Cryosurgery of genital warts in cases in which podophyllin treatment failed or was contraindicated. Br J Vener Dis 1977;53:49–53.
10 Bergman A, Bhakia NN, Broen EM: Cryotherapy for treatment of genital condyloma during pregnancy. J Reprod Med 1984;29:432–435.
11 Benedet JL, Miller DM, Nickerson KG, Anderson GH: The results of cryosurgical treatment of cervical intraepithelial neoplasia at one, five, and ten years. Am J Obstet Gynecol 1987;157:268–273.
12 Jensen SL: Comparison of podophyllin application with simple surgical excision in clearance and recurrence of perianal condylomata acuminata. Lancet 1985;ii:1146–1148.
13 Simmons PD, Langlet F, Thin RN: Cryotherapy versus electrocautery in the treatment of genital warts. Br J Vener Dis 1981;57:273–274.
14 Murphy M, Bloom GD: Podophyllin or podophyllotoxin as treatment for condylomata acuminata? Papillomavirus Rep 1991;2:87–91.
15 Beutner KR, Von Krogh G: Current status of podophyllotoxin for the treatment of genital warts. Semin Dermatol 1990;9:148–151.
16 Simmons PD: Podophyllin 10% and 25% in the treatment of ano-genital warts: A comparative double-blind study. Br J Vener Dis 1981;57:208–209.
17 Beutner KR, Conant MA, Friedman-Kien AE, Illeman M, Artman NN, Thisted RA, King DH: Patient-applied podofilox for treatment of genital warts. Lancet 1989;1:831–834.
18 Lassus A, Haukka K, Forsstrom S: Podophyllotoxin for treatment of genital warts in males: A comparison with conventional podophyllin therapy. Eur J Sex Transm Dis 1984;2:31–33.
19 Krebs HB: Treatment of genital condylomata with topical 5-fluorouracil. Dermatol Clin 1991;9:333–341.
20 Chan S, Middleton RK: Bleomycin treatment of warts. DICP 1990;24:952–953.
21 Marker T, Reymann F: Treatment of plantar warts with cytostatics. Acta Derm Venereol (Stockh) 1967;47:359–362.
22 Steinberg BM, Abramson AL, Horowitz B: Effects of alpha-interferon on cultured laryngeal papilloma cells; in Howley P, Broker T (eds): Papillomaviruses: Molecular and Clinical Aspects. New York, Liss, 1985, pp 413–420.
23 Auborn KJ, Steinberg BM: Therapy of papillomavirus-induced lesions; in Pfister H (ed): Papillomaviruses and Human Cancer. Boca Raton, CRC, 1990, pp 203–223.
24 Senff H, Reinel D, Matthies C, Witts D: Topical 5-fluorouracil solution in the treatment of warts – Clinical experience and percutaneous absorption. Br J Dermatol 1988;118:409–414.
25 Krebs HB: Prophylactic topical 5-fluorouracil following treatment of HPV-associated lesions of the vulva and vagina. Obstet Gynecol 1986;64:773–778.
26 Sillman FH, Sedlis A: Anogenital papillomavirus infection adn neoplasia in immunodeficient women: An update. Dermatol Clin 1991;9:353–369.
27 Chiang CM, Ustav M, Stenlund A, Ho TF, Broker TR, Chow LT: Viral E1 and E2 proteins support replication of homologous and heterologous papillomaviral origins. Proc Natl Acad Sci USA 1992;89:5799–5803.
28 Ustav M, Stenlund A: Transient replication of BPV-1 requires two viral polypeptides encoded by the E1 and E2 open reading frames. EMBO J 1991;10:449–457.
29 Cowsert LM, Munger K, Howley PM, Baker CC: The E6-E7 region of HPV-16 is sufficient for altered differentiation of human keratinocytes; in Howley P, Broker T (eds): Papillomaviruses: Proceeding of a UCLA Symposium. New York, Wiley-Liss, 1990, pp 265–271.
30 Munger K, Phelps WC, Bubb V, Howley PM, Schlegel R: The E6 and E7 genes of the human papillomavirus type 16 together are necessary and sufficient for transformation of primary human keratinocytes. J Virol 1989;63:4417–4421.
31 Sterling J, Stanley M, Gatward G, Minson T: Production of human papillomavirus type 16 virions in a keratinocyte cell line. J Virol 1990;64:6305–6307.
32 Meyers C, Frattini MG, Hudson JB, Laimins LA: Biosynthesis of human papillomavirus from a continuous cell line upon epithelial differentiation. Science 1992;257:971–973.
33 Howett MK, Kreider JW, Cockley KD: Human xenografts: A model system for human papillomavirus infection. Intervirology 1990;31:109–115.
34 Bonnez W, Rose RC, Da Rin C, Borkhuis C, de Mesy-Jensen KL, Reichman RC: Propagation of human papillomavirus type 11 in human xenografts using the severe combined immunodeficiency (SCID) mouse and comparison to the nude mouse model. Virology 1993;197:455–458.
35 Swinehart J, Skinner R, Miller BH, McCarty M, Eron L, Kraus S, Shrum J, Beutner K, Korey A, Rudd R, Willson C, Dito J, Song J, Lu SP, Orenberg E: Intralesional chemotherapy with MPI 5003 for treatment of condylomata acuminata. Am Acad Dermatol, 52nd Annu Meet, Washington, Dec 1993.
36 Shikowitz MJ, Steinberg BM, Galli RL, Abramson AL: Histological and molecular analysis of cottontail rabbit papillomavirus-induced papillomas treated with hematoporphyrin derivative photodynamic therapy. Arch Otolaryngol Head Neck Surg 1988;114:175–178.
37 Abramson AL, Shikowitz MJ, Mullooly VM, Steinberg BM, Amella CA, Rothstein HR: Clinical effects of photodynamic therapy on recurrent laryngeal papillomas. Arch Otolaryngol Head Neck Surg 1992;118:25–29.
38 von Knebel Doeberitz M, Oltersdorf T, Schwarz E, Gissmann L: Correlation of modified human papilloma virus early gene expression with altered growth properties in C4-1 cervical carcinoma cells. Cancer Res 1988;48:3780–3786.
39 von Knebel Doeberitz M, Rittmüller C, zur Hausen H, Durst M: Inhibition of tumorigenicity of cervical cancer cells in nude mice by HPV E6-E7 anti-sense RNA (letter). Int J Cancer 1992;51:831–834.
40 Steele C, Cowsert LM, Shillitoe EJ: Effects of human papillomavirus type 18-specific antisense oligonucleotides on the transformed phenotype of human carcinoma cell lines. Cancer Res 1993;53:2330–2337.
41 Storey A, Oates D, Banks L, Crawford L, Crook T: Anti-sense phosphorothioate oligonucleotides have both specific and non-specific effects on cells containing human papillomavirus type 16. Nucleic Acids Res 1991;19:4109–4114.
42 Lappalainen K, Urtti A, Jaaskelainen I, Syrjanen K, Syrjanen S: Cationic liposomes mediated delivery of antisense oligonucleotides targeted to HPV E7 mRNA in CaSki cells. Antiviral Res 1994;23:119–130.
43 Smotkin D, Prokoph H, Wettstein FO: Oncogenic and nononcogenic human genital papillomaviruses generate the E7 mRNA by different mechanisms. J Virol 1989;63:1441–1447.
44 Del Vecchio AM, Romanczuk H, Howley PM, Baker CC: Transient replication of human papillomavirus DNAs. J Virol 1992;66:5949–5958.
45 Cowsert LM, Fox MC, Zon G, Mirabelli CK: In vitro evaluation of phosphorothioate oligonucleotides targeted to the E2 mRNA of papillomavirus: Potential treatment for genital warts. Antimicrob Agents Chemother 1993;37:171–177.
46 Thierry F, Yaniv M: The BPV1-E2 trans-acting protein can be either an activator or a repressor of the HPV18 regulatory region. EMBO J 1987;6:3391–3397.
47 Crooke ST: Progress in oligonucleotide therapeutics. Am Soc Pharmacol Exp Ther, Spring Meet, Anaheim, Apr 1994.